5/98

FIRE ON EARTH

FIRE ON EARTH

DOOMSDAY, DINOSAURS, AND HUMANKIND

John and Mary Gribbin

St. Martin's Press
New York

Library of Congress Cataloging-in-Publication Data

Gribbin, John R.
    Fire on Earth : doomsday, dinosaurs, and humankind / by John and
Mary Gribbin.
        p. cm.
    ISBN 0-312-14335-4 (alk. paper)
    1. Asteroids. 2. Comets. 3. Catastrophes (Geology). I. Gribbin,
Mary. II. Title.
QB377.G83   1996
551.3'9—dc20   96-6071
                                                                        CIP

First published in Great Britain by Simon & Schuster

First U.S. Edition: July 1996

10 9 8 7 6 5 4 3 2 1

[The] story of how Phaëton, child of the sun, harnessed his father's chariot, but was unable to guide it along his father's course and so burnt up things on the earth and was himself destroyed by a thunderbolt, is a mythical version of the truth that there is at long intervals a variation in the course of the heavenly bodies and a consequent widespread destruction by fire of things on the earth.

Plato
Timaeus

Wait for the day when the sky will pour down visible smoke, enveloping all men: a dreadful scourge. Then they will say: 'Lord, lift up the scourge from us. We are now believers.'

Koran, 44: 1

ACKNOWLEDGEMENTS

Many people involved in research into the origins of the kind of 'fire on Earth' referred to by Plato helped in the preparation of this book by discussing their work with us and supplying copies of their relevant publications. In reverse alphabetical order, we would particularly like to thank: Iwan Williams, of Queen Mary & Westfield College, London; Chandra Wickramasinghe and Max Wallis, of University College, Cardiff; Duncan Steel, of the Anglo-Australian Observatory; Michael Rampino, of New York University; Brian Marsden, of the Harvard-Smithsonian Center for Astrophysics; Bill McCrea, of the University of Sussex; Neil McBride, of the University of Kent at Canterbury; Jane Luu, of the Harvard-Smithsonian Center for Astrophysics; David Hughes, of Sheffield University; Iain Gilmour, of the Open University; Victor Clube, of the University of Oxford; Mark Bailey, of Liverpool John Moores University; and David Asher, of the Anglo-Australian Observatory.

INTRODUCTION

Interest in the possibility that objects from space – comets or meteorites – might strike the Earth is higher than it has ever been, partly as a result of the collision of fragments from comet Shoemaker-Levy with Jupiter. The story even made the cover of *Time* magazine in the summer of 1994. But as well as attracting popular interest, this has become a subject taken increasingly seriously by scientists in the past few years.

Part of the reason for this is the now well-established theory that the 'death of the dinosaurs' was caused by the impact of a large meteorite with the Earth. The crater produced by this event has now been identified in the Yucatan peninsula of Mexico. The evidence for this impact, and the consequences, has given credibility to an idea that has been around for some time, suggesting that lesser impacts – on the scale of the famous Tunguska event – occurred more frequently in the recent history of the Earth than they do today, and played a part in shaping the development of human civilization through their effects on the environment, up to and including triggering ice ages.

Almost twenty years ago, geologists discovered a thin layer of unusual debris in rocks 65 million years old from around the world. The debris had all the hallmarks of extraterrestrial origin, and it had been laid down at exactly the time when some catastrophe had caused the death of the dinosaurs. This led to a theory that a giant meteor had struck the Earth at that time, starting huge forest fires and spreading dust high into the air, from where it had settled down to form this unusual layer. But where was the direct evidence for such an impact?

At the beginning of the 1990s, an investigation of a crater, 180 km across, buried under the town of Chicxulub on the northern coast of the Yucatan peninsula in Mexico, found the 'smoking gun' that shows the impact theory is correct. Rock samples collected during the search for oil in the Gulf of Mexico showed both the extent of the feature and its age. There is no doubt that a large object from space impacted with the Earth at this spot at the time the dinosaurs died out and the extraterrestrial sediments were laid down. And there is no reasonable doubt that the three events were linked.

Obviously, impacts on this scale are rare. But there have been other impacts over the eons, and other 'extinctions' of life on Earth. What are the implications for human civilization?

The 'Tunguska Event' provides a striking demonstration of terrestrial life's vulnerability to quite small cosmic impacts. On 30 June 1908, a fragment of comet exploded over Siberia, releasing a fireball with 2,000 times the force of the Hiroshima atomic bomb. It felled trees over an area of more than 2,000 square km, and started fires that burnt out a region half that size. Tremors produced by the impact were recorded in St Petersburg 4,000 km away, and around the world. And yet, this devastating event was caused by a fragment of ice weighing only about 100,000 tonnes, vapourising in the atmosphere as it burnt up at an altitude of about 6 km. This is a tiny fraction of the size of the Yucatan meteorite, and it didn't even reach the ground.

But it was only by luck that the explosion ocurred over the Siberian forests, and not a little further west, over the populated part of Russia – perhaps even over St Petersburg, which would have had broad implications for world history, since Lenin was living there at the time.

Some theorists argue that the Tunguska Event may have been caused by a fragment of leftover material from an interplanetary stream which has already destroyed civilizations twice. If this hypothesis is correct, the world is in for another bout of fire from the heavens in about a thousand years' time.

Astronomers know that there are objects like very large cometary nuclei, lumps of icy matter more than 200 km across, in the outer part of the Solar System. One example is the object Chiron, orbiting beyond Jupiter. It seems likely that one comet in a thousand is such a giant, and that most of the mass of the cloud of comets that surrounds the Solar System is in the form of giants.

Gravitational disturbances, chiefly caused by Jupiter and Saturn, will perturb such objects so that about once in every 200,000 years a giant falls in to the inner part of the Solar System. Computer simulations show that the orbits the comets end up in are not stable, but are influenced by chaotic dynamics. They must, however, end up in a Sun-grazing orbit, and get broken up into pieces by the Sun in the same way that Shoemaker-Levy has been broken into pieces by Jupiter. The result is a stream of debris orbiting round the Sun and crossing the orbit of the Earth.

When the Earth passes through such a stream of material, it will collect a rain of fine dust over a timescale of thousands of years, until this fine material is blown away by the solar wind. As the Earth sweeps up about 570 million tonnes of dust each year, with a certain amount settling out of the atmosphere, the resulting 'load' of material in the air will average out at about a thousand billion kilograms. This is sufficient to reduce surface temperatures (by acting as a sun-shield) by 3–5 degrees Kelvin, perhaps triggering an Ice Age.

But this is not the only problem. Mixed in with the dust stream, and still there when all the fine dust has blown away, there will be fragments of comet ranging from a few centimetres to a few tens of kilometres across. The impact that wiped out the dinosaurs was caused by an object only about 10 km in diameter, but the probability of a direct hit by such a large fragment is less than that for smaller fragments.

We are, it is claimed, living in the aftermath of the break-up of a giant comet in the inner Solar System. This event may have been associated with the most recent Ice Age, which began about 100,000 years ago. It produced a stream of Sun-orbiting material linked with the Taurid meteor stream, which peaks around 30 June in daylight hours but is visible as 'shooting stars' in the night skies of November.

The Earth passes through the thickest part of this belt of debris every 3,000 years, and there is evidence that this happened most recently in AD 500 and before that in 2,500 BC. On both occasions, Tunguska-like events would have been common, with one impact in each region the size of England over a period of a hundred years or so.

This could explain, as we shall propose later, the collapse of some ancient civilizations, the 'Dark Ages' of

Europe, and recurring legends about fire from the skies. The Tunguska Event itself came at just the right time of year to fit the pattern, as an isolated straggler in the stream, but the next main date to watch out for is the year 3000, give or take 200 years. For once, the scientists involved are happy that they will not be here to test their prediction.

The extent to which the possibility of future impacts is taken seriously is, however, indicated by the existence of projects such as Spacewatch and Spaceguard, to look for threatening objects and (one day soon) to establish the capability of changing their orbits using 'Star Wars' technology.

That, in outline, is the story we have to tell in this book. It is no surprise that it makes headlines in popular newspapers and the cover of *Time* magazine, and features in TV specials. But how well does it stand up to closer scrutiny? What evidence is there that the evolution of life on Earth has indeed been moulded by impacts from above? Where's the beef? If you want to know more, read on.

We begin our story with the evidence that it really was a bolt from the heavens that finished off the dinosaurs. We end it with some thoughts about how we might prevent a similar catastrophic event bringing an end to our own civilization. And in between, we intend to provide enough scientific beef to convince you that there is a lot more to all this than sensational headlines in popular newspapers.

John Gribbin
Mary Gribbin
July 1995

Table 1:
Planetary Properties

Planet	Period (year)	average distance from Sun	size	mass	atmosphere	moons	rings
Mercury	0.24	0.4	0.4	0.06	n	0	no
Venus	0.62	0.7	0.9	0.8	y	0	no
Earth	1	1.0	1.0	1.0	y	1	no
Mars	1.88	1.5	0.5	0.1	y	2	no
Jupiter	11.8	5.2	11.2	318	y	16+	yes
Saturn	29.5	9.5	9.5	95	y	18+	yes
Uranus	84	19.2	4.0	14.5	y	15+	yes
Neptune	165	30.1	3.9	17.2	y	8+	yes
{Pluto	249	39.5	0.18	0.003	?	1	no}

Notes

1 All measurements are in units where the Earth's properties are defined as equal to 1. Period = duration of full orbit of Sun; i.e. Earth period = 1 year

2 Pluto is not really a planet, but for historical reasons is usually included in this list.

3 In the 'Gap' between Mars and Jupiter lies the asteroid belt, most of them between 1.7 and 4 times as far from the Sun as we are. They have periods of between 3 and 6 years. The Kuiper Belt of comets lies beyond Neptune, between about 35 and 1,000 times as far from the Sun as we are.

CONTENTS

FIRE ON EARTH

1

THE DEATH OF THE DINOSAURS

The era of the dinosaurs came to an end 65 million years ago, when the Earth was struck by an object only about 10 km across, but which imparted a blow so severe that it disrupted the environment and hurled so much material into the air that when it settled out it formed a layer of clay still detectable in sedimentary rocks around the world. It was the equivalent of a rock the size of Mount Everest, travelling ten times faster than the fastest bullet, producing an impact so severe that the entire Earth shifted in its orbit by a few dozen metres and the length of the year changed by a few hundredths of a second.

Massive interest in the possibility of such objects from space striking the Earth was roused in the early 1980s, when the discovery of the signature of extraterrestrial material, found in these ancient clays, was announced. The link between this particular impact and the death of the dinosaurs struck a chord in the popular imagination, and encouraged an explosion of scientific interest in the implications. Since then an overwhelming weight of evidence has confirmed that the clays were produced by the aftermath of an impact from space, and the final confirmation of the theory came in the early 1990s with the identification of the spot on the Earth where the impact from space occurred – the 'smoking gun', in the form of a crater the right age, the right size, and in the right place to do the job.

But although it was the discovery of the cosmic cause of the 'terminal Cretaceous event', as it is known, that made the investigation of impacts from space a growth industry among astronomers, geologists and paleontologists in the 1980s, a handful of scientists had long been aware of the hazards which lumps of rock and ice orbiting in the inner Solar System pose for life on Earth. We shall discuss the origins of some of this interstellar debris later in this book; what matters is that there are pieces of material ranging in size from dust grains to boulders to flying mountains; and that the orbits of some of them around the Sun cross the orbit of the Earth, giving rise to the possibility of an occasional cosmic collision.

The rocky lumps of debris are known as asteroids or meteoroids. When a small grain of such material from space reaches the Earth's atmosphere, it burns up in the fiery streak known as a meteor. If it is large enough to survive the journey through the atmosphere and strike the surface of the Earth it is called a meteorite (by extension, objects from space that strike the surfaces of other planets, or moons, are also known as meteorites). The icier 'space mountains' lose material by evaporation as their orbits take them near to the heat of the Sun, spewing out great streamers of gas which reflect the light from the Sun. They are known as comets, but there is little real difference between the core of a comet and an asteroid. Many of the meteoroids that cross the orbit of the Earth are probably

2

leftover pieces from comets that have had too many close approaches to the Sun and have lost all their icy material. As far as the effects on life on Earth are concerned, it would make little difference whether the surface of our planet were struck by a flying mountain made of rock or by an equivalent mountain made of ice and rocks. All that matters is the mass of the object, and the speed with which it hits the surface of the Earth.

The first person who really spelt out the implications of such an impact for life on Earth was a British schoolteacher, Joe Enever, who published a detailed description of the effects of such an impact in the mid-1960s. The inspiration for Enever's calculations was the evidence that the Earth bears of such impacts in the distant past – evidence in the form of scars such as the famous Barringer Crater in Arizona. The energy involved in producing these scars was colossal, far greater than the energy of any nuclear explosion produced by humankind. It comes from the combination of the mass (m) of the speeding object and the object's speed itself (v) known as kinetic energy, and given by the simple expression: $E = \frac{1}{2} mv^2$.

When a car is stopped by its brakes, the kinetic energy is turned into heat and the brakes warm up; when a lump of rock 10 km across hits the Earth and is brought to an abrupt halt, a great deal more kinetic energy is converted into heat, with explosive results.

Enever chose as his case study a large impact feature in South Africa called the Vredevoort Ring. Putting in the numbers appropriate for this ring – a feature about 140 km across, formed some two thousand million years ago – he found that the energy which produced the original crater must have been equivalent to around a thousand million

3

megatons of TNT. For comparison, a mere 10 megaton explosion would be spectacularly large for a nuclear device today. An impact big enough to produce the Vredevoort Ring would require a lump of rock about 10 km across to strike the Earth, arriving at a collision speed of 50 km per second, or more; and from what they know about the number of asteroids and comets in the Solar System and the nature of their orbits, astronomers calculate that such a collision ought to occur somewhere on Earth roughly once every hundred million years.

It happens that there are fewer known craters on the surface of the Earth than this estimate of the frequency of such collisions would imply. But this is no surprise, for two reasons. First, about two thirds of the surface of our planet is covered by water, and any meteorites that fell in the sea would leave no visible scars today. Secondly, even on land the natural processes of erosion smooth out old impact scars, which are also covered by vegetation. They can be very hard to detect from the surface and a true picture of just how battered the face of our planet is has only emerged in the 'space age', with the aid of satellite photography and other techniques for detecting large geological structures from orbit. This, of course, is another reason for the upsurge of interest in impacts from space in recent decades.

You might think that a meteorite striking the sea would have less impact than one striking the land, that the collision would be damped down by the cushioning layer of water. Unfortunately that isn't so.

One of the most surprising things about our planet, to anyone who has sat on the beach and watched the ocean breakers roll in, but has not been told how shallow the waters of the ocean are, is how insignificant a puddle even the greatest ocean is.

In relative terms, oceans really are very shallow. The North Atlantic, for example, averages a depth of 3 km and is about 4,800 km wide. In order to have the same proportions, a puddle a mere 3 mm deep would have to be nearly 5 metres across. Even the floors of the deepest ocean trenches lie only 11 km below the surface of the sea, and the average depth of the ocean is just 3.7 km. A ball of rock 10 km in diameter could sit on the floor of the ocean with more than half of its bulk above the waves!

The thinness of the life-zone on Earth also applies in the other direction – upwards. Even the atmosphere of the Earth – at least, the breathable part in which life exists – extends no more than about 10 km above our heads. (Mount Everest rises nearly 9 km above sea level, and breathing is pretty difficult on its summit.)

Some ecologists have graphically described the zone in which life can exist on Earth (oceans and atmosphere put together) as a 'thin green smear' on the surface of a planet which is itself nearly 13,000 km across. The power of this image is rammed home by the realisation that the distance from the top of the tallest mountain on Earth to the bottom of the deepest ocean trench is only about 20 km: a distance which, horizontally, you could cover in about 10 minutes by car, and would represent less than a gentle day's walking. If the Earth were reduced to a ball 13 cm across (about the size of a grapefruit), the 'green smear' would be a film just 0.2 mm thick over the surface of the ball. At a little over 12 miles, the vertical distance from the bottom of the deepest oceanic trench to the top of the tallest mountain corresponds to a bit less than a half-marathon, a distance thousands of people run each year for fun, without realising that in a couple of hours they have run the same distance as from the bottom of the sea to the top of Mount Everest. Put in those terms it is easy to see how thin the green smear really is, and begin to understand why it can be so dramatically disrupted by cosmic impacts.

To an object 10 km across, coming in at a speed of 50 km per second, or more, the waters of the sea would provide no more of a brake than the atmosphere. The meteorite would strike with almost undiminished force on the thin oceanic crust, liberating its kinetic energy in a thousand-million megaton explosion and punching a hole right through to the molten magma (the stuff that spews out of volcanoes) below the crust. Since an oceanic impact is most likely, that was the kind Enever focused on for his detailed calculations. Water rushing in to the resulting fiery pit, about 150 km across, would eventually quench the magma, covering it with a scar tissue of new crust, but not before 16,000 cubic km of water had been vapourized, blanketing the Earth in clouds. The cloud cover, Enever pointed out some 30 years ago, would reflect away the Sun's heat, cooling the world below so much that the water vapour (enough to cover the entire globe with the equivalent of 5 cm of rain) would freeze and fall as snow, killing the plants on which animals depend for food.

Why didn't the calculation create a sensation in the 1960s? Perhaps because the time wasn't ripe. Perhaps because, intriguing though the conclusions were, Enever had no direct, hard evidence that such a global disaster had occurred at the time of the death of the dinosaurs. Or perhaps it was because, since this was only a hobby of his, he published his calculations in the form of an entertaining article in the magazine *Analog* (better known for its science fiction, although it does also publish articles of speculative science fact, like Enever's contribution). But later studies have shown just how close he was to hitting the mark.

The relative obscurity of Enever's article (buried in a journal 'only' read by a few hundred thousand SF fans, instead of one read by a few thousand paleontologists) is certainly not the whole reason for its failure to set the scientific world on fire, because several scientists published related ideas in the 1970s and were almost entirely ignored by their peers.

There have been several other major catastrophes in the history of the Earth, known as extinctions, in which many species of plants and animals died out entirely. And there have also been many more lesser events, in which fewer species suffered, but enough went to provide geologists and paleontologists with a marker they can use to divide up the geological timescale. In some cases it was mainly land-based life forms that suffered; in other cases, ocean dwellers were the hardest hit. And on a few occasions death seems to have struck indiscriminately at life forms both on land and in the sea.

The geological timescale is, indeed, all about extinctions. Looking back to successively older layers of rock (which, by and large, means looking deeper below the surface), geologists find rocks in which different species are common. The species found in the oldest rocks give way, fairly abruptly, to different species in higher layers of rock. The transition zones between the two kinds of ancient populations are generally much narrower than the zones in which one kind of fauna and flora flourish.

Many causes have been suggested for these changes. One of the most likely is that the way the continents move about the surface of the Earth has changed the geography and climate from time to time; so species that have flourished under one regime find themselves suffering a change they cannot cope with – drought or flood, perhaps, or an Ice Age, or global warming – and die out, to be replaced by species that are better adapted to the new conditions. Some of these extinctions may well have been caused by the impacts of objects from space, although that possibility has only been taken seriously by most scientists since the 1980s.

Whatever the cause of these dramatic events (and they need not all have the same cause), on a geological time-scale they have to be seen as part of the natural environment of the Earth. This is important. Some people mistakenly interpret the evidence for such dramatic changes as being in some sense 'anti-Darwinian', suggesting that evolution does not, after all, proceed by the action of natural selection winnowing the variety of individual members of a species to ensure that only those best suited to their environment (best 'fitted', like the fit of a key in a lock) survive. But if Charles Darwin had been aware of the pattern of great extinctions, and the way in which life on Earth has recovered after each of them, he would surely have seen these events, and their aftermaths, as evidence in support of his theory.

What seems to happen is that many species become better fitted to their environments in succeeding generations; some become perfectly adapted in the way that the long, curved beak of a humming bird is adapted to suck nectar from a particular species of flower. When a catastrophe occurs, the superbly adapted species die out, and new species evolve from the survivors – in exactly the way that Darwin surmised – to take advantage of the new environmental conditions. In this sense, extinctions may encourage evolution, by 'winnowing out' long-established species and providing scope for newcomers (and we may be the direct beneficiaries of this process); but there is nothing going on here that cannot be directly explained in Darwinian terms.

There have been five really massive extinctions of life on Earth since the earliest fish evolved. These, along with smaller extinctions and other geological events, are used by scientists to divide the geological timescale into periods. The first major extinction occurred about 438 million years (Myr) ago. This defines the boundary between what geologists have termed the Ordovician and the Silurian periods. Another great extinction occurred just over 360 Myr ago, marking the end of the Devonian period; a third (the greatest of them all) about 250 Myr ago, at the end of the Permian. Great extinction 4 occurred 215 Myr ago, at the end of the Triassic period. The most recent occurred 65 Myr ago, closing the Cretaceous period, and wiped out, among other things, the dinosaurs. Together, these events are sometimes referred to as the 'Big Five'. This doesn't mean there were no extinctions – even widespread extinctions – in the intervals between the Big Five. The end of the Jurassic period, for example, some 145 Myr ago is defined by minor extinctions. But the Big Five were, as their name implies, the biggest.

The period after the Cretaceous is known as the Tertiary, so the extinction that involved the death of the dinosaurs defines the Cretaceous–Tertiary boundary. This is usually abbreviated K–T, rather than C–T, to

avoid confusion with the Cambrian period. Some of the
greatest extinctions definitely coincide with great geological events, such as the formation of the 'supercontinent'
of Pangea, and its breakup about 215 Myr ago. But others,
including the K–T extinction, do not correspond to any
comparably dramatic changes in geography. As long ago
as 1970, Digby McLaren used his Presidential Address to
the Paleontological Society to put forward the argument
that at least one of the extinctions, near the end of the
Devonian, was caused by the impact of 'a large or very
large meteorite' in the ocean. He specified a watery impact
site because this particular extinction singled out certain
kinds of marine organisms for death, and he suggested
that:

> Presumably on impact with the ocean surface or at
> a certain depth below the surface, the missile will
> explode with an enormous release of energy . . .
> The turbulence of the tidal wave and accompany-
> ing wind, followed by the gigantic runoff from land
> would induce a turbid environment far longer than
> could be survived by bottom-dwelling filter feeders
> and their larvae.[1]

Clearly, McLaren was not an avid reader of *Analog*, or he
would have realised just how modest his proposal was. But
in spite of its modesty, and in spite of the highly visible
platform from which the proposal was made, few paleontologists took much notice of the idea. In the 1970s, it was
left to astronomers to make further speculations about the
effects of cosmic impacts on the Earth.

This followed a well-established tradition, but one
entirely unknown to geologists and paleontologists.
Back in 1958, the astronomer Ernst Öpik, an Estonian

comet expert who was then dividing his time between Ireland and the United States, published a paper in the *Irish Astronomical Journal* under the title 'On the catastrophic effects of collisions with celestial bodies'.[2] This was just one of at least a dozen papers he wrote on the subject, all published in astronomical journals and largely ignored. The American, Harold Urey – who had won the Nobel Prize for chemistry in 1934, worked on the atomic bomb project and later became interested in the origin of the planets – published his contribution on the cosmic impact theme under the title 'Cometary collisions and geological periods' in *Nature* in 1973.[3] Urey's model was very fully worked out, including a suggestion that the shock waves from impacts might trigger outbursts of volcanic activity, and jumping off from the idea that glassy fragments known as tektites (of which more later) are both associated with geological boundaries and produced by cometary impacts. A few years later, the British astronomer Victor Clube was asking (in print) the question 'Does our Galaxy have a violent history?' and providing an answer that included, in part, the possibility of life-threatening impacts of meteorites.[4] But the thing that made people sit up and take notice of the idea of a link between cosmic impacts and mass extinctions was the discovery of an enormous concentration (relatively speaking) of the rare metal iridium in a thin layer of clay that marks the boundary between the Cretaceous and Tertiary periods.

Before we look at the details, though, there are a couple of important points to take on board. The first is that although the K–T boundary event is often referred to as the time of 'the death of the dinosaurs', the dinosaurs were by no means the only species that

11

suffered. At the end of the Cretaceous, 70 per cent of all species alive on Earth were wiped out. Both plants and animals, on land and in the sea were affected; this was one of the greatest disasters that has ever hit our planet. The second is that the dinosaurs themselves were, by 65 million years ago, highly successful and long-established in a variety of ecological niches. There were dinosaur grazers and dinosaur carnivores, dinosaurs that flew in the sky and dinosaurs that swam in the seas. If you look at the variety of mammals around on Earth today, from cows to tigers, bats to dolphins, there was a dinosaur that filled the corresponding ecological niche in the late Cretaceous, and then some. This was not a variety of life form in terminal decline, ready to keel over and give up the ghost if the environment twitched slightly. Whatever did for the dinosaurs was as all-pervading a disaster as you would need to kill off all the mammals alive on Earth today.

Just as the era of the dinosaurs ended with a major catastrophe, so it began with an even bigger extinction that disrupted life on Earth at the end of the Permian period, about 250 Myr ago. This was such a major event that geologists use the marker it provides in the fossil record to denote not just the end of a period but the end of an era of geological time – the Paleozoic. There have only been three eras in the past 590 million years (the interval since the Precambrian, a largely blank area in the history of our planet which covers the first seven-eighths of its existence, but which has left few fossils for study). The Palaeozoic era ran from 590 Myr ago to 248 Myr ago, and saw the development of fish in the sea, the emergence of life on to the land, and the evolution of reptiles. After the great terminal Permian extinction, dinosaurs evolved and, from about 230 Myr ago, flourished, emerging in one period of geological time (the Triassic) and prospering during the next two periods (the Jurassic and the Cretaceous) which, together with the Triassic, make up the Mesozoic era. Like the terminal Permian catastrophe, the K–T event was so severe that it marks the end of an era, and the last 65 Myr is known as the Cenozoic era.

The dinosaurs flourished for roughly 150 million years, from the end of the Triassic, about 213 Myr ago, to the end of the Cretaceous, 65 Myr ago. For comparison, the earliest human ancestor that is distinct from the common ancestors we share with other apes was around no more than five million years ago. Stretching the definition of people to its utmost, we have been around for a little more than 3 per cent of the time that the dinosaurs dominated the Earth.

Even then, the dinosaurs never completely disappeared. A few species survived the K–T catastrophe, and some of their descendants are still around today, in the form of birds. It is intriguing that there is just a hint here that dinosaurs which had evolved feathers, giving a warm layer of insulation, may have been better able to survive the catastrophe at the end of the Cretaceous than their cousins. But that is another story, beyond the scope of the present book. The big question that still baffled geologists and paleontologists at the end of the 1970s was just why such a huge variety of successful and long-established species were wiped out at the end of the Cretaceous. Was it a result of climatic change? Was there some environmental catastrophe? Or what?

One of the biggest questions was just how long it had really taken for the catastrophe to occur. A thin layer of rock, representing the transition from the Cretaceous to the Tertiary, might really represent a band of sediments laid down over hundreds of thousands of years, and even then it is hard to prove that a particular species was alive and well right up to the transition zone. The consensus among the experts was that dinosaurs and other Cretaceous life forms had gone into a gradual decline over millions of years, before being replaced – for whatever reason – by new variations on the evolutionary theme.

It was with this in mind that a team of researchers headed by Luis Alvarez, of the Lawrence Berkeley Laboratory, in California, set out to analyse the layer of clay seen clearly at the K–T boundary in the rocks near the town of Gubbio, in northern Italy. The clay is part of a sequence of sediments that had originally been laid down in the sea, but had been lifted up by geological activity over millions of years and is now easily accessible on dry land. It can be seen, in a gorge just outside Gubbio, as a thin red-brown stripe less than a centimetre thick, between two layers of limestone. The limestone below the red stripe contains Cretaceous fossils; the limestone above, Tertiary fossils. The K–T boundary is by no means a subtle geological marker – you can 'see the join' without even getting out of your car. But what made the join, and how long did it take to do the job?

The layer of clay is so thin that if it had been laid down by normal geological processes it would have been produced in less than a thousand years. This is like one tick of the clock on the geological timescale, and suggests that, by geological standards, the switch from Cretaceous to Tertiary conditions happened instantaneously. Nobody really believed that the change had happened overnight, but it would be interesting to find out exactly how long it had taken for the clay layer to be deposited, and Alvarez approached the puzzle from a new perspective.

He was a physicist and, like Urey, he worked on the Manhattan Project. He also invented the radar-controlled 'blind landing' system for aircraft, and in 1968 he won the Nobel Prize for his contributions to elementary particle physics. He became interested in geology (especially the geology of the K-T boundary) through his son, Walter, a Professor of Geology at the University of California at Berkeley. Walter wanted to know how long it had taken for the Gubbio clay layer (about as thick as a pencil) to be deposited. Presumably it corresponded to the transition from Cretaceous conditions to Tertiary conditions; but had that transition taken place over ten thousand years, a hundred thousand years, or something in between?

Luis and Walter Alvarez discussed how the timescale might be measured, and hit upon the idea of measuring the amount of iridium in the clay. Iridium (which is a member of the same family of metals as gold and platinum) is very rare on the surface of the Earth, but tiny quantities are constantly falling on to our planet from space, as part of a rain of microscopic dust grains (micrometeorites). If iridium comes down steadily, at the same rate per hundred years, throughout geological time, then maybe by measuring the amount of iridium in the clay they would be able to tell how many hundred years it had taken for the clay layer to form.

It would be no easy task to measure accurately the tiny traces of iridium involved, but Luis Alvarez knew just the people who could do it. They were Frank Asaro and Helen Michel, nuclear chemists at the Lawrence Berkeley Laboratory, who were experts in the techniques required to identify small quantities of this kind of material. But when the analysis was carried out, it produced a completely unexpected result. In the first measurements of the iridium content of the Gubbio clay, the team found 30 times more iridium than there was in a comparable thickness of rock from either the Cretaceous sediments (just below the clay layer) or the Tertiary sediments (just above the clay layer). In fact, their technique had been overloaded; more accurate measurements soon showed that there is actually about 300 times more iridium in the layer than there ought to be.

This was the opposite of what they had expected to find. They had expected to see a small amount of iridium, showing that the clay had been laid down over a small span of time, less than a thousand years. A layer of clay like the Gubbio layer, about a centimetre thick, would normally take at the most about 10,000 years to be laid down. In order to contain so much iridium, if the layer had formed over a timescale long enough to accumulate the iridium from the steady rain from space it would have been almost *four million* years in the making. This made a complete nonsense of their hopes of estimating the timescale of the transition in terms of the steady rain of iridium from space. On that basis, it would have meant that this tiny layer had been laid down over an enormous span of geological time, while nothing else went on on Earth to disturb it, and evolution was held in suspended animation. The only logical explanation of the iridium anomaly was that some great catastrophe had spread iridium around the globe, and had killed off the dinosaurs and their contemporaries in the process. But what?

After careful deliberation, Luis Alvarez and his colleagues concluded that the most likely candidate was the impact of a meteorite some 10 km across with the Earth. Iridium is rare in the crust of the Earth because, being a heavy element, it settled down into the molten iron core of our planet when the world was young and hot. But there are relatively large concentrations of iridium (compared with the concentrations in the surface rocks of the Earth) in meteorites. A sufficiently large meteorite impact could have spread fine dust, debris from the meteorite itself and the pulverised rock from the crater blasted out of the Earth's crust, in a cloud around the Earth that slowly settled out to become the clay layer marking the K-T boundary – a clay layer enriched with iridium from the meteorite. And, the team argued, that shroud of dust around the Earth would block out the sunlight, killing plants that depended on the Sun for photosynthesis and plunging the planet into an instant Ice Age.

It was the publication of this variation on the cosmic impact theme, in the journal *Science* in June 1980,[5] that, at last, made people sit up and take notice of the idea. It struck a chord in the popular imagination, and, among other things, led to the development of the idea of a 'nuclear winter' following all-out nuclear war (indeed, the K-T impact scenario was sometimes referred to as 'impact winter'). But it also received a lot of criticism from many geologists and paleontologists, understandably reluctant to accept that a couple of physicists and chemists muscling in on their territory had solved one of the greatest mysteries in the fossil record.

But the great thing about the new theory was that, in the best scientific tradition, it made predictions that could be tested. If the Alvarez team was right, there ought to be a layer of iridium-enriched clay at the K-T

17

boundary just about everywhere on Earth; there ought to be other signs of events associated with a major impact 65 million years ago; and there might well be a crater the right age, yet to be discovered – although nobody counted any chickens on that one, since the odds were 2:1 that the impact had occurred in the sea. But one of the strongest pieces of circumstantial evidence that the theory had going for it was that an impact big enough to wreak the kind of destruction the team was talking about, involving an object 10 km across, ought to occur about once every hundred million years. Indeed, the once-in-a-hundred-million-year interval between such events is intriguingly similar to the rate at which the great extinctions have occurred on Earth, although nobody, in 1980, wanted to make too much of that coincidence.

Since 1980, an overwhelming weight of evidence has accumulated to confirm the reality of the blow from space that caused the terminal Cretaceous event – in some cases, the evidence even comes from studies carried out in the hope of proving the hypothesis wrong. For example, an early criticism of the idea was that perhaps the dinosaurs and their contemporaries really had faded out slowly, over ten million years or more. This wouldn't completely rule out the meteorite scenario, but it would reduce the impact to playing the role of a 'last straw', finishing off populations already in decline. So paleontologists scrutinised the layers of rock above and below the K–T boundary carefully, to see just when the fossils of various species disappeared from the record. The classic fossils from before the boundary are the ammonites, the spiral fossils that every schoolchild knows. Before these new studies, it had been thought that the ammonites faded out well below the K–T boundary; but recent studies have shown that they are present right up to the boundary, but nowhere above it. The same applies to dinosaur remains themselves (although dinosaur remains are nowhere common in the rocks, and it involved a great deal of painstaking work to establish that they really were still flourishing up to the geological moment when the meteorite struck).

It wasn't just animals that were affected. After all, the reason why the K–T boundary was identified in the fossil record in the first place is because it is an obvious 'biomass killing horizon', in the paleontological jargon. The remains of plants on land (and their fossilised seeds and pollen) and of plankton in the oceans tell the same story – one kind of life flourishing in the Cretaceous, a sudden decline in the number of species right on the K–T boundary, and then a recovery, with new forms of life evolving in the Tertiary. The more closely the paleontologists looked at the fossil record, the more clear it became just how sharp that boundary was.

There was also plenty of evidence of the physical impact of the K–T event. The 'Gubbio clay' has been found at many sites around the world, and anomalously high amounts of iridium – in many cases a thousand times the normal concentration – have been found associated with the K–T boundary at more than a hundred sites around the globe. Tiny diamonds have also been found in the K–T boundary clay in Alberta, Canada. The diamonds really are tiny, specks of material that can only be seen with the aid of a microscope, and at one time it was thought that they probably came in with the object that hit the Earth 65 myr ago, rather than being formed in the impact itself. If that were the case, they would be independent proof that extraterrestrial material arrived on Earth in large quantities at the time the dinosaurs died. But more detailed analysis of the chemical composition of the diamonds shows that it is much more likely that they were produced in the impact itself; their composition does not match that of other tiny diamonds found in meteorites, but they are similar to diamonds produced in large explosions of TNT.

Even stronger proof that there was a big impact at that time comes in the form of tiny grains of so-called shocked quartz found in the boundary layer material at many sights. This kind of material, only produced by extreme pressures, had previously been found only at the sites of known impact craters and in the craters produced by nuclear explosions. The grains can only have been produced in the meteorite impact at the K–T boundary, and spread around the world in the plume of debris from the explosion. Another form of shocked material, called stishovite, has also been found in the K–T boundary clay from a site in New Mexico. And then there are the tektites. These are little glassy drops of material, often associated with meteorite craters, formed in the heat of the impact and sprayed out over a wide area. Tektites come in a variety of shapes – round, teardrop and dumb-bell shapes are common – and may be a few millimetres across. They are found in plenty in the K–T boundary layer across North America. These tektites, however, are no longer in their original glassy form, but have been converted into a kind of clay by chemical processes over the tens of millions of years that have elapsed since the impact. Baffled to find clay 'tektites' in the boundary layer, geologists were finally persuaded of their meteoritic origins when little blobs of material were found in the boundary layer on Haiti – spherules which had an outer layer of clay but still contained an inner core of glass formed in the heat of the original impact.

Some geologists fought a rearguard action against the impact hypothesis, arguing that the kind of evidence being turned up by the renewed global interest in the K–T boundary in the 1980s could all be explained in terms of volcanic activity. They pointed out that there had been a massive episode of volcanic activity in what is now India, producing huge sheets of volcanic material known as the Deccan Traps, at about the right geological time. This may well have had a widespread environmental impact. But although some volcanoes do emit traces of iridium, most do not; and no volcano produces shocked material like the quartz grains and stishovite found in the boundary layer, or tektites.

The clinching blow to the volcano hypothesis came when the supporters of the impact theory quickly realised that a very good way to trigger a massive burst of volcanic activity would be to hit the Earth with a large meteorite. The shock waves from such a blow might well have triggered activity at weak spots in the Earth's crust. This explains the 'coincidence' of the volcanic activity with the meteorite impact, and added to the difficulty of life at the end of the Cretaceous.

One thing you certainly don't get out of volcanic eruptions is the kind of complex chemical compounds known as amino acids. But there are even traces of amino acids at the K–T boundary layer – 31 different amino acids in all, 18 of which have been found nowhere else on Earth, but all of which are known to be present in interplanetary material such as meteorites.

By the early 1990s, there was no reasonable doubt that a cosmic impact really had been responsible for the death of the dinosaurs. Nobody knew exactly how the extinctions had occurred; the original idea of darkness and cold enveloping the Earth had begun to look too simplistic, but there was a wealth (almost an embarrassment) of detailed alternative suggestions, ranging from acid rain to a great overheating of the world caused by an enhancement of the greenhouse effect. But this was regarded as a matter of detail, no real embarrassment to the impact theory. The one fact that still niggled proponents of the impact theory, and which gave comfort to the diehards who tried to deny the mounting weight of evidence, was that nobody had yet found a good candidate for the crater that such an impact ought to have produced. Where was the smoking gun?

At the end of 1990, it was found. A crater the right size, in the right place, and the right age to explain the terminal Cretaceous event. There was no longer any room to doubt that the dinosaurs had died as a result of a blow from space, and the extra information provided by the discovery of the impact site, together with other new evidence gleaned in the 1990s, gave a good indication of how and why they had died.

Notes

1 *Journal of Paleontology*, volume 44, p. 801 (1970).
2 Volume 5, p. 34.
3 Volume 242, p. 32. As well as being prescient, Urey's paper is short, clear and entertaining. Referring to comet impacts, he says that 'I have also suggested that the geological periods were terminated by such collisions, but this was published in the *Saturday Review of Literature*, and no scientist except me, as far

as I know, reads that magazine.' This may have been the first suggestion of a connection between events like the death of the dinosaurs and cometary impacts.

4 *Vistas in Astronomy*, volume 22, p. 77.
5 Volume 280, p. 1,095.

11

THE SMOKING GUN

Some of the evidence for the impact at the end of the Cretaceous was already beginning to pin down the location of the impact by the early 1990s. Diamonds from Alberta, tektites from across North America, shocked quartz and stishovite, and even more impressive tektites from Haiti (a layer of them half a metre thick, also rich in iridium and shocked quartz) were helping to persuade many of the experts involved in the hunt for the smoking gun that they should concentrate their attention in and around North America itself, either on the land mass or in the Caribbean Basin.

One of the first good candidates for the impact site had been found on the North American continent itself, in the form of the Manson impact structure in Iowa. The ancient crater turned out to be exactly the right age, 65 Myr, and to be ideally located to explain the concentration of shocked quartz grains found in the K–T layer in North America. But it was too small – only 35 km across, when the experts were calling for a crater at least 150 km in diameter to explain the terminal Cretaceous event. And it was on land, which was puzzling, because by the end of the 1980s more detailed analysis of the K–T boundary layer had shown clearly that it contained material recycled from rocks characteristic of the sea floor. And yet, the chance of such a meteorite striking the Earth exactly at the time of the death of the dinosaurs is only 4 per cent – one chance in 25. Could it be a coincidence? Or could there have been two (or more!) impacts that occurred at the same time not by chance, but because they were caused by meteoroids that had been following the same orbit through space, alongside one another?

This is by no means a crazy idea. The impact of the Shoemaker-Levy comet with Jupiter in 1994 showed exactly what could happen, and bore out the earlier calculations of astronomers concerning a similar disaster affecting the Earth. Shoemaker-Levy was broken up into a 'string of pearls' by a close encounter with Jupiter, and on their next orbit the 'pearls' slammed into Jupiter, one after the other, in quick succession. Calculations carried out in the 1980s showed that a comet could be broken up by a close passage near the Sun (or, indeed, by a close encounter with Jupiter), and that the fragments could then strike the Earth, like the blast of a shotgun, producing a scattering of craters, all the same age, and combining to produce the extinctions that mark the end of the Cretaceous. It began to look as if perhaps there was no single smoking gun after all. But in 1988, the first clear evidence of a truly dramatic single impact was found.

The discovery was made by geologists Joanne Bourgeois, Thor Hansen, Patricia Wiberg and Erle Kauffmann at sites near the Brazos River, in Texas. They found a thick layer of unusual sediments, which they explained as having been produced by the effects of a massive tidal wave (a tsunami), immediately underneath a layer of sediment enriched with iridium, at the K–T boundary.

The obvious explanation was that the sought-for meteorite had splashed down in the Caribbean region, sending the tsunami racing across the nearby shallow seas (at that time, this part of what is now Texas was covered by coastal waters) and producing the disturbed sediments. The iridium thrown up by the impact then settled down out of the air on top of the disturbed layer.

The evidence, which the team published in the journal *Science* in July 1988, is truly impressive.[1] The unusual sediments form a band of sandstone up to 130 cm thick, which contains pieces of shells, bits of wood, fish teeth and clumps of mudstone up to a metre across. The highly disturbed, jumbled layer gradually gives way to smoother layers of sandstone on top, with traces of ripples caused by waves. More ordinary mudstone layers above and below the jumbled sandstone contain the remains of creatures that live at depths around 75 metres, so it seems that the sandstone must have been laid down in that depth of water, offshore. The lumps of mudstone mixed in with the sandstone are so big, though, that it would take a current moving at a speed of a metre per second to fling them about and mix them in with the other material; but the layer of fine sandstone at the top of the disturbed layer is so thin that it could have been laid down in a single day.

All the evidence points to a tsunami with a wave height greater than 75 metres, ripping across what is now the Gulf of Mexico, temporarily exposing the sea bed and racing up on to the nearby land, scouring up the material that would form the sandstone bed as it went and then carrying the material back on to the continental shelf as the waters retreated. Significantly, although in theory a huge volcanic eruption, like that of Krakatau, could produce such a tsunami in its immediate locality, there is no trace of volcanic ash from the fallout from such an explosive eruption in the sandstone bed.

Similar sediments were already known from around the Gulf of Mexico and the Caribbean Basin, but had traditionally been interpreted in terms of changes in sea level, taking place over much longer periods of time than a single day. When these other sediments were re-examined in the light of the new evidence, they soon confirmed the picture painted by Bourgeois and her colleagues. Walter Alvarez and several of his colleagues, for example, studied a site at the Arroyo el Mimbra, in Mexico, where geologists looking for oil in the 1930s had reported an unusual sandstone bed. They found that the layer, which had also been formed underwater 65 Myr ago, could be read like a book to tell the story of the impact.

At the bottom of the sandstone bed, there was a layer of clay spherules, some of which turned out to have glassy cores – tektites thrown out by the meteorite impact, which had landed on the sea floor while it was laid bare by the rush of the tsunami up the nearby land. The jumbled sandstone above, similar to the Texas layer, showed the mixed up debris dumped by the backwash from the tsunami. And at the top of the jumbled layer a fine sandstone layer showed traces of the ripples produced as waves continued to rock back and forth across the region in the hours that followed. Finally, the icing on the cake: the whole thing is topped off by a thin layer enriched with iridium – the K–T boundary itself. Similar markers of the passage of the tsunami are found from Alabama to Guatemala, as well as on the island of Haiti.

Slightly more conservative than Enever, Alvarez calculated that all the damage could have been done by the impact of an object about 10 km across releasing energy equivalent to the explosion of about a hundred million megatons of TNT. That would still be equivalent to ten thousand times the energy that would have been liberated by the simultaneous explosion of all the nuclear weapons on Earth at the height of the cold war, in the 1980s, before the superpowers started to dismantle their stockpiles.

So where is the crater? An impact large enough to have caused all this activity should have left a sizeable scar on what was then the shallow sea floor. And since a lot of the sea floor has since been raised up by geological processes to make dry land, there was even a good chance that all or part of the crater might be on land today. And then, just when the academic geologists were beginning to zero in on the site of the impact, they found that it had already been discovered by their industrial counterparts – that, indeed, it had been identified in the 1970s, before the Alvarez team had even discovered the iridium layer. The reason nobody in academic circles knew about it was that the discovery had been made by geologists working in the oil exploration business: their bosses had been reluctant to spill the beans about the interesting geology of the Yucatan Peninsula in Mexico, for fear of giving valuable free information to their competitors.

The buried crater had been discovered in 1978 by Glen Penfield, a geophysicist working for the Western Geophysical Company of Houston, Texas. Western Geophysical had been hired by the Mexican national oil company, Pemex (Petróleos Mexicanos), to carry out a survey of the Yucatan area from the air, measuring the pattern of magnetism in the rocks. The magnetic survey showed a huge semicircular pattern, a feature which had no right being there. His curiosity roused, Penfield dug out of the Pemex files a geological map of the region that had been made back in the 1960s, using gravity measurements. Such a gravity map highlights regions where the rocks have different densities – and it clearly showed the trace of the other half of the magnetic semicircle. Putting the two maps together, Penfield found that he was looking at a perfectly circular feature, a bulls eye 180 km across, centred on the coastal village of Puerto Chicxulub. A keen amateur astronomer, he immediately realised that he was looking at a buried impact crater. But in 1978, nobody was looking for impact craters in the region, and there had been no public fuss about the idea that a meteorite impact may have caused the death of the dinosaurs. Besides, Pemex wouldn't allow the data to be released.

In 1981, after the discovery of the iridium layer at the K–T boundary and the first wave of publicity for the Alvarez team's idea, Penfield was allowed to mention the findings in the Yucatan to a meeting of the Society of Exploration Geophysicists. (By then, Pemex were sure there was no oil in the region, so they weren't too concerned about secrecy.) Nobody took much notice – by a bizarre twist of fate, most of the cratering experts in the United States were attending a specialist meeting on impacts that week, in another city, and although the discovery of the Chicxulub crater was now officially public knowledge, it still hadn't reached the notice of the people who were by now beginning to look for just such a crater.

Penfield carried on investigating the structure as best he could. Pemex had drilled several exploratory wells in the Yucatan region since the 1950s, and these had found a thick layer of andesite rock about 1.3 km below the surface. Andesite is just the kind of material that you would expect to find in a meteorite crater; rocks produced from the re-solidified molten material formed in the heat of the impact. Unfortunately, the original cores drilled from these wells had been lost track of – for a time, it was thought that they had been destroyed in a warehouse fire but happily this turned out not to be the case. So, in trying to find out more about the buried structure, Penfield had to make do with bits and pieces of core left over in rubbish heaps at the original drilling sites.

This was the state of play early in 1990, when Alan Hildebrand, of the University of Arizona, visited Beloc, in southern Haiti, to study the jumbled tsunami debris there. As well as the traces of the tsunami, he found large quantities of material that had been ejected from the impact site – tektites, shocked quartz, and all the rest. There was so much of this ejected material, and some of the fragments were so big (tektites 8 mm across), that he calculated that the impact site must be no more than 1,000 km away, to the south. But where?

The latest Haiti discoveries were reported at the annual Lunar and Planetary Science Conference in Houston in March 1990. Afterwards, a local reporter, Carlos Byers from the *Houston Chronicle*, approached Hildebrand. He told him that way back in 1981 he had heard a Houston-based geophysicist called Glen Penfield talking about a huge buried crater at Chicxulub, in the Yucatan . . .

Hildebrand and Penfield immediately joined forces for a renewed assault on the geology of the region. They soon found shocked quartz in some of the material from the old drilling sites. Since then, much more evidence has been gathered (including the re-discovered original cores from the old drilling programme) to show that the buried feature is indeed an impact crater 65 million years old. It isn't obvious from the surface, because it is buried under a layer of limestone laid down in the shallow waters of the region before geological activity lifted what is now the Yucatan Peninsula up and made it dry land. Limestone is made from the carbonate shells of marine creatures that die and fall to the bottom of the sea, and there is an equivalent layer of limestone underneath the andesite of the crater itself, showing that the impacting object really did strike in a shallow, offshore sea – a fact which, as we shall soon explain, may have been particularly unfortunate for the dinosaurs and their contemporaries. The Yucatan impact site is also, perhaps coincidentally, on almost exactly the opposite side of the Earth from the site of the volcanic activity that produced the Deccan Traps; shock waves rippling outward from the impact would have tended to focus together again in just about that part of the world.

The last piece of evidence came when standard dating techniques involving measurements of radioactive argon in the rocks showed that the crater had formed exactly 65 Myr ago, and the same techniques gave the same age for the tektites from Beloc. New drilling is being planned to flesh out the details, but the evidence is already overwhelming. There really was an impact big enough to wreak global havoc in the shallow sea just south of what is now North America at the end of the Cretaceous – an impact which left one of the largest impact craters, some 180 km in diameter, yet identified on the surface of the Earth.

But it wasn't alone. The Manson impact structure was formed at the same time, and geologists have also found another crater 65 Myr old: a multi-ringed impact site 100 km across at Popigai, in Siberia. Were the dinosaurs actually wiped out by a *triple* whammy? It looks very likely. Indeed, this may be a conservative assessment, since there are several other features of about the right age (but not yet dated accurately) in a line linking these sites. They include an impact structure 20 km across at Avak in Alaska, a pair of craters at Kara in western Siberia, and others. To eyes that have seen the photographs of the impact of the pieces of Shoemaker-Levy with Jupiter, this is highly intriguing.

But let's be cautious, and concentrate on the three well-dated simultaneous impacts. If they were produced by fragments of a comet that broke up as it passed near the Sun, astronomers can even infer the time of day and the season in which the impacts occurred. For the three pieces of interplanetary debris, travelling nearly together through space on their way out from the Sun, to strike simultaneously in Yucatan, Iowa and northern Siberia, it must have been summer in the northern hemisphere, with the north polar regions tilted towards the Sun and offering an inviting target. And since the pieces of comet were coming directly from the direction of the Sun, striking the side of the Earth facing the Sun, the impact must have occurred about midday, probably out of a clear blue summer sky. There is even independent confirmation of the time of year when disaster struck, in the form of plants in a fossilised lily pond in Wyoming, which have been dated to the K–T boundary and which died at the stage of growth corresponding to early June. Three simultaneous impacts would be ample to explain the terminal Cretaceous event; and the experts are now getting a pretty clear idea of the lethal cocktail of consequences that actually did the killing.

The shallow sea of what is now the Yucatan Peninsula was one of the worst possible places for the meteorite to strike. The sea floor in that part of the world was rich in a variety of sediments laid down in shallow water which had dried out and reflooded several times over the preceding interval of geological time. This included not only the carbonates of the limestone, but also other material known (for obvious reasons) as evaporites – extensive beds of material such as anhydrite, up to a kilometre thick and rich in sulphur, and calcite, another form of carbonate. In the heat of the impact, huge quantities of both sulphur dioxide and carbon dioxide would have been released from these rocks. Water rushing back in to the hot spot and being blasted into steam (as in Joe Enever's scenario) would have combined with the sulphur dioxide to produce huge quantities of sulphuric acid. The sulphuric acid droplets, carried high into the atmosphere by the blast and heat, would first act as a very efficient sunshield (forming a high altitude haze beneath which the Earth would cool close to freezing point for several decades), and then fall out as extremely acidic rain: bad news for any form of life on the surface of the Earth, and especially bad news for sea creatures such as ammonites, whose shells would be eaten away by the acid.

Then, as the sulphuric acid cleared from the atmosphere over a timescale of a few years, the carbon dioxide would begin to work, boosting the greenhouse effect. As if that were not enough, water vapour (produced in copious quantities in the impact by the Enever effect) is a very effective greenhouse gas, and would add to the global warming. Overall, this would heat the surface of the Earth by as much as 10 °C for at least 500,000 years. All this because the impact site was rich in evaporites – a kind of rock which is found only over about half of one per cent of the surface of the globe! And on top of that, the dinosaurs and their contemporaries had to cope with consequences that would have followed wherever the meteorites had struck.

The first problem was heat and fire. Particles thrown out by the impact, including the material that formed the tektites, would have flown on ballistic trajectories, re-entering the atmosphere like smaller meteorites in their own right, carrying kinetic energy which had to be released as heat as they were brought to a halt. According to calculations made by H. J. Melosh and colleagues at the University of Arizona, the blast could actually have stripped away some of the Earth's atmosphere into space; while the energy of the particles re-entering what was left of the atmosphere would have produced a heating of 10 kilowatts per square metre at the Earth's surface for several hours after the impact – a heating level which Melosh has graphically described as 'comparable to that of a domestic oven set at "broil"'.[2] Many of the dinosaurs may literally have been roasted alive, while this level of heating of the globe helps to explain one of the most striking features of the K–T boundary, alongside the iridium layer – a layer of soot produced by a global firestorm of almost unimaginable extent.

The soot layer has been found at many sites around the world, as far apart as Europe and New Zealand, and its composition and origins have been studied by several groups of researchers, including Iain Gilmour (now of the Open University in Britain) and his colleagues. The location of the soot in the boundary layer clay shows that the fires had started even before all of the ejected material from the impact(s) had settled, and chemical analysis shows that the layer resembles charcoal produced by forest fires. These studies are pretty sophisticated. They don't just tell you that there is soot present at the K–T boundary, but what kind of soot. And they don't just tell you that the soot came from forest fires but, in some cases at least, what kind of trees were burning. The presence of a compound known as retene in the K–T boundary clay at several sites shows that many of the burning trees were conifers, since this compound is produced by combustion of resins found in pine trees.

One of the most intriguing pieces of evidence confirming that smoky fires were the origin of the soot is the presence of compounds known as fullerenes in the layer. It has long been known that pure carbon comes in two forms – a very tough crystalline form, called diamond, in which each carbon atom is locked in position relative to its neighbours, and a softer form, known as graphite, in which the carbon atoms are arranged in layers that slide easily over one another. It was only in the 1980s that chemists found a third form of pure carbon, in which sheets of carbon are rolled up, naturally, to form spheres, or tubes. It was dubbed 'buckminsterfullerene' after Buckminster Fuller, who designed the geodesic domes (almost exactly the shape of a soccer ball) which spherical fullerenes resemble. It is now known that fullerenes are commonly produced in smoky flames, and some such fullerenes are found in the soot from the K–T boundary, where they had lain for 65 million years, even though Nobody on Earth knew that such a form of carbon could exist until years after the Alvarez team came up with the idea that a meteorite had produced the iridium at that boundary.

The most likely scenario for the fires at the end of the Cretaceous is that trees were flattened by the powerful winds (at least 500 km per hour) produced by the impact, dried by the kind of heating described by Melosh's team, and ignited either by the heat arising from the impact or by lightning activity sparked by the settling dust in the atmosphere. The fires produced 70 billion tonnes of soot, corresponding to the combustion of 25 per cent of the organic material estimated to have been present on Earth at the time. The 'smoking gun' really did produce smoke, with a vengeance. The direct effects of the fires alone would include: a period of darkness and cold caused by the smoke (for comparison, a relatively modest forest fire in northern California in September 1987 reduced temperatures in a smoke-filled valley by 15 °C for a week); the spread of poisons such as nitrous oxide and mutagens produced in the fires around the globe; and an enhancement of the greenhouse effect as the carbon dioxide from the burned material increased the concentration of this greenhouse gas in the atmosphere to at least three times its present day concentration (the soot itself, of course, represents only the fraction of the original biomass that *didn't* burn completely to carbon dioxide).

Overall, Gilmour and his colleagues have identified a dozen 'environmental stresses' caused by the K–T impact.[3] The strong winds and tsunamis lasted for a matter of hours; fires lasted for months, as did the darkness and cold partly caused by the fires; the greenhouse effect began to take a grip as the darkness cleared, boosted early on by the presence of water vapour in the air, and maintained by the long-term presence of carbon dioxide; poisons and mutagens remained active for years, as did the effects of acid rain; the ozone layer must have been severely disrupted by the disturbance to the atmosphere; and then there was the volcanic activity triggered by the impact.

The variety of these stresses is important, because they affect different species in different ways, and to different degrees. When the meteorite impact idea was first widely discussed, in the early 1980s, some critics of the theory argued that the pattern of extinctions seen at the K–T boundary could not be explained by a single cause. But that now seems to be an argument in favour of the meteorite impact idea, since it is hard to see how any other event could have produced such a complex mixture of environmental stresses all operating at the same time. What kind of a disaster could affect species as different as the dinosaurs and the ammonites, while allowing crocodiles to survive? The answer seems to be that they were different, but related, disasters.

On land, animals that depended on green plants for their food, such as the plant-eating dinosaurs, suffered when the plants (those that hadn't burned in the global fire) were deprived of sunlight for photosynthesis and died (and meat-eating dinosaurs suffered, of course, when the plant-eating dinosaurs died out). In the sea, acid rain proved catastrophic. And some species, or enough individual members of those species, were able to survive in less badly affected regions of the globe and spread out again as conditions improved, because they were just plain lucky. Some even benefited because of their way of life; food chains based on scavenging, or feeding off the detritus that washed down into rivers and lakes, provided food in plentiful supply for creatures that could make use of it in the cold, dark months before the green plants saw the Sun and began to grow again.

But what was a disaster for 70 per cent of all species on Earth was an opportunity for the 30 per cent who survived. As we described in another book, *Being Human*, it was precisely because the dinosaurs, in particular, suffered the catastrophe of the K–T event that the ancestors of all modern mammals (including ourselves), were able to diversify and spread out to fill the niches previously occupied by dinosaurs. Those ancestors of ours were small creatures that kept well out of the way of large dinosaurs and probably lived off insects (bound to survive the K–T event – as somebody once quipped, the most likely survivors of all out nuclear war would be cockroaches) and perhaps by scavenging scraps of dead dinosaur (in plentiful supply just after the meteorite struck). It seems as certain as anything studied at such a distance can be that if the entire geological history of the Earth were altered by removing only one event, the K–T impact, then we would not be here. Students working on the problem with Walter Alvarez came up with a graphic analogy, comparing the opportunities that were suddenly made available to mammals at the start of the Tertiary with the excellent job prospects there would be for people with fresh PhDs in geophysics if 70 per cent of all the professors of geophysics were fired overnight.

The way in which the survivors spread out and adapted to take advantage of the ecological niches that were now open to them exactly matches the Darwinian description of evolution at work. Indeed, the image of small numbers of survivors evolving in different ways to fit different ecological niches closely resembles the classic example of Darwinian evolution used by Charles Darwin himself. On the Galapagos Islands, where a few individual birds arrived and settled, descendants evolved and diversified into many different varieties of finch to fit the different ecological conditions prevailing on different islands. Evolution theory cannot tell you how and when the physical conditions affecting life will change – whether there will be a new Ice Age, or a drought, or fire falling from the sky – but it can tell you how species that survive the change will adapt to the new conditions. Catastrophism is in no sense 'anti-Darwinian'; catastrophes are simply part of the natural environment within which Darwinian evolution operates.

In a sense, therefore, we should be grateful for the K–T event. But we should also be just a little in fear of it. It simply represents the most extreme example of a kind of cosmic impact which, recent research has shown, is much more common than used to be thought. Impacts on the Chicxulub scale may only occur every hundred million years or so, and the Chicxulub impact itself, for the reasons we have just explained, may have been a particularly disastrous example of its kind. But tiny meteors, grains of cosmic dust that burn up in the atmosphere, strike the Earth every day, and cosmic rubble exists on all scales between these two extremes, so impacts on all scales between these two extremes are possible. As a rough rule of thumb, the smaller the impact, the more frequently such events occur. And an impact can be a *lot* smaller than the K–T event and still do a great deal of damage to the terrestrial environment. Without looking at anything on the scale of the impact that killed off the dinosaurs, it would not take much of an impact to cause severe disruption to human life on Earth. The death of the dinosaurs (which led to the rise of the mammals and ultimately of ourselves) is only the beginning of our story. From now on, we shall be looking at events that directly affect human beings – starting with a near miss that came within a few minutes of altering the course of twentieth-century history.

Notes

1 Volume 241, p. 567.
2 *Nature*, volume 343, p. 251 (1990).
3 Geological Society of America, Special Paper number 247, 1990; also in *Global Catastrophes in Earth History*, edited by V. L. Sharpton & P. D. Ward.

III

A CLOSE ENCOUNTER – OR TWO!

Cosmic impacts on the scale of the disaster that killed off the dinosaurs are certainly rare by human standards. There hasn't been one in the past 65 million years, and the rule of thumb is that they occur every hundred million years or so. But that still means that there must have been more than 40 events on this scale since life emerged on Earth about four billion years ago – not just 40, because impacts of all kinds must, as we shall see, have been more common when the Earth was young and the Solar System was littered with cosmic debris.

It is impossible to imagine what such an impact must have been like. Scientists can make the calculations and give you numbers for the size of the explosion and so on, but even they cannot really picture the event those numbers describe. But for the past half-century or so, every educated person has had an image of the most powerful kind of destruction caused by human activities, the explosion of a nuclear bomb. And it is now known that an explosion which occurred over a sparsely populated region of Siberia in 1908 had a destructive power roughly equal to that of the largest explosion most of us can imagine – one of the larger hydrogen bombs developed during the cold war. This explosion, roughly as big as the largest explosion ever produced by human activities, was caused not by the impact of a large asteroid with the Earth, but by the break-up in the air over Siberia of just a small chunk of wayward cometary material, a fragment of its parent body.

The explosion took place near the Podkammenaya Tunguska River, in a wild part of the taiga north of Lake Baikal. Just after seven o'clock in the morning, on 30 June 1908, a bright, pale blue fireball streaked across the sky from southeast to northwest, leaving a thick trail of dust in the atmosphere. The fireball was described as blindingly bright, making the sun look faint by comparison. It was caused by a fragment of cosmic debris which entered the Earth's atmosphere somewhere over western China, and travelled on a shallow downward trajectory across the sky and down towards the ground, trailing a series of loud explosions (described as 'thunderclaps', and possibly sonic booms) in its wake. It exploded just before 7.15 am, still about 6 km above the ground, at a spot which has since been pinpointed as 60° 55′ North, 101° 57′ East. Eye witness reports describe a column of fire and smoke which rose 20 km into the air, and was visible from hundreds of kilometers away.

Almost 900 km away to the south, scientists at the Irkutsk Magnetic and Meteorological Station recorded seismic disturbances caused by the blast, beginning at 7.18 am and lasting for more than an hour. While the seismic detectors were still recording these disturbances, around 8.00 am the meteorological instruments recorded a fluctuation in atmospheric pressure, as the blast wave from the explosion reached Irkutsk. The spreading blast wave was soon recorded on instruments in Europe, including meteorological stations in Berlin and London. The seismic shock was recorded 4,000 km away, in St Petersburg, and at other sites around the world, while at Potsdam, in Germany, a second fluctuation in atmospheric pressure was recorded the following day: a trace of the 'other side' of the Tunguska blast wave, which had travelled right around the world and arrived in Potsdam from the opposite direction.

The explosion, known as the 'Tunguska Event', devastated the forest over an area of more than 2,000 square km, with a central region about half that size set ablaze by the heat. Estimates of the size of the explosion vary widely, up to an extravagant suggestion that it was equivalent to the explosion of 100 megatonnes of TNT. About 20 megatonnes, roughly a thousand times as powerful as the nuclear blast that destroyed Hiroshima in 1945, would be a conservative estimate.

But there was no nuclear blast at Tunguska in 1908.[1] All you need to account for the devastation (as in the case of the K–T event) is good old kinetic energy – in this case, the energy associated with a chunk of material weighing only a few million tonnes, moving at a speed of only about 30 km per second relative to the Earth. As far as the kinetic energy, and its consequences, are concerned, it doesn't matter what that 100,000 tonnes was made of. It could have been ice, or rock, or a lump of metal – except that a lump of metal would have survived the passage through the air and hit the ground, leaving a crater. Indeed, 100,000 tonnes of green cheese hitting the Earth at 30 km per second would do just as much damage. At these speeds, the kinetic energy in each gram of any material is roughly the same as the chemical energy released in the explosion of a gram of TNT.

The Tunguska explosion was definitely an airburst – shown by the pattern of fallen trees under the centre of the explosion – suggesting that the object that exploded was more friable than a lump of metal (though stronger than a lump of cheese) and literally could not stand the heat of its own passage through the air. On the other hand, the lump of cosmic debris did get down to 6 km from the surface, after a long passage through the air on a slanting trajectory. That probably rules out a lump of ice, which would not have lasted that long. So the best contender for the cause of the Tunguska Event is a lump of rocky material, perhaps 100 metres across. This is what would traditionally be called an asteroid or a meteor (technically, not a meteorite since it didn't hit the ground). Comets contain large chunks of rock and icy material, frozen together in a lump, and like many other asteroids, this particular chunk of rock may well have come from the break-up of a cometary nucleus.

Disintegration of a rocky object also explains one of the strangest aftermaths of the Tunguska Event. That night, 30 June, and the following night, 1 July, people across Europe and North America were astonished by the brightness of the night sky. It was, to be sure, high summer, when the northern hemisphere evenings stay light until late. But nobody could remember it staying so light, for so long into the night, even at the end of June. In Sweden and Scotland, photographers seized the opportunity to take photographs by the available natural light at midnight, using exposures of about a minute (the same light ruined astronomical photographs taken at the observatory in Heidelberg after midnight). The light after midnight was widely described as being sufficient to read a newspaper by; and, characteristically, cricketers in England took the opportunity to indulge in midnight games. On 30 July, the *New York Times* described 'remarkable lights' that had been 'observed in the northern heavens', while in Belgium the great red glow seen in the sky after sunset was likened to the light from a huge, distant fire. The lightness of the night sky, which gradually faded over the following days, and all the other spectacular effects resulted from fine particles of dust from the asteroid, spread high in the atmosphere by the explosion and reflecting the Sun's light from below the horizon. It is also possible that some of the dust came from a small cometary tail associated with the object. This tail, pushed by the Sun's light, would have been pointing ahead of the object as it hit the atmosphere of the Earth, spreading fine dust further to the north and west.

The influence of the dust in the upper atmosphere was also shown in a series of observations which, by a happy coincidence, had just begun in California. In May 1908, the American astronomer C. G. Abbot had begun observations aimed at determining exactly how much heat from the Sun arrives at the surface of the Earth, and how much the Sun emits (the 'solar constant'). These measurements involved very accurate observations of the transparency of the air at different wavelengths – they showed a slight decrease in the transparency of the atmosphere from about two weeks after the Tunguska Event until after the middle of August. (At the time, of course, Abbot knew nothing about this explosion, which had happened in an almost uninhabited region on the other side of the world.) There was no equivalent decrease in transparency in the same months in other years. With hindsight, it is clear that dust from the Tunguska Event was blocking out some of the light from the Sun, minutely decreasing the temperature at the surface of the Earth for a few weeks, as far away as California.

Where did the dust go? Some of it fell on the Antarctic ice cap, where traces of dust from layers of ice the right age have been recovered and analysed. The composition of these dust grains closely matches that of tiny (sub-milli-metre sized) metallic spheres found near the site of the Tunguska explosion, showing that they came from the same cosmic source. And, just in case you are wondering, yes, both the spherules found at the Tunguska site and the 'Tunguska dust' from Antarctica do contain relatively large amounts of (among other things) iridium.

None of this was known until long after the events of 1908. It took a long time for rumours and stories about fire from the sky to filter out of the backwoods and attract the attention of scientists in European Russia, let alone the rest of the world. By the time scientists were thinking about going to Siberia to investigate what had occurred there, Russia was involved in World War I, then the Revolution, and then the Civil War. It was not until well into the 1920s that any serious scientific investigation of the site was possible, and even then the scientists laboured against extreme handicaps.

The story has been told in great detail by E. L. Krinov, in his book *Giant Meteorites*.[2] To give you a feel for how different things were in those days from today, it's worth remembering that it was only after the Revolution in 1917 that Russia started using the same calendar as the rest of Europe. The date we have given for the Tunguska Event, 30 June 1908, is the date according to the modern Gregorian calendar. But in Russia at that time, the date was actually 17 June, on the old Julian calendar still in use there. The discrepancy between these calendars also explains, of course, why the Bolshevik Revolution in 1917 is widely referred to as the 'October Revolution', even though it took place on 7 November according to the modern calendar.

Astronomers don't have much trouble with the date of the event, since there was only one Tunguska explosion in the summer of 1908, and eyewitness accounts are clearly referring to the same thing, whether they are dated 17 June or 30 June.

Astronomers have to be more careful with the exact time of day given in different accounts of the explosion. It happened a little after midnight GMT – the best timings come from the seismic records, and set the exact moment as 14 minutes and 28 seconds after midnight. But what time was it at the site of the explosion? In terms of time zones, that part of Siberia is seven hours ahead of Greenwich, so it was just after 7.14 am. But that is the time for the centre of the zone, and the site of the explosion is so far from the centre of the zone that the precise local time, by the Sun, was 7.02 (and 16 seconds) in the morning at the moment the fireball exploded. The difference is important, because astronomers are keenly interested in the exact direction in which the fireball was travelling relative to the Sun. This provides information about what orbit the rock was in before it reached the atmosphere of the Earth.[3] The general conclusion is that the rock was coming from the direction of the rising Sun. One eye witness gave a graphic description of a fiery streak crossing the sky like a piece of glowing material 'broken off' from the Sun and hurtling down to Earth. Whatever it was, in order to create that optical illusion it certainly must have come in on an orbit from the direction of the Sun!

But what we are interested in now is not the exact origin of this piece of cosmic debris, but the effects it produced in the region where it exploded. Remember that this was a relatively small piece of rock, maybe 100 metres across – only one hundredth of the diameter of the object that landed near what is now the Yucatan Peninsula 65 Myr ago – and since mass is proportional to volume, which goes as radius cubed, maybe one millionth of the mass of that object. If the two objects had hit the Earth with roughly the same speed (which can only be pure guesswork), then the kinetic energy turned into heat in the Tunguska Event must have been about one millionth of the energy released in the terminal Cretaceous event.

The effects of that release of energy were dramatic and, had they occurred near any major centre of population on Earth, the history of the twentieth century might well have taken a different course. Indeed, as several people have pointed out, the site of the explosion lies almost due east of St Petersburg. If the Tunguska rock had arrived on the same trajectory through space a few hours later, when the Earth had turned a little more and it was just after 7 am solar time in that city, St Petersburg would have been destroyed, along with most of its inhabitants – which included, on 30 June 1908, a certain Vladimir Ilich Ulyanov–Lenin.

The scientist who put the study of the Tunguska Event on a scientific footing was Leonid Alexeivich Kulik, who was also the first person to point out how close St Petersburg came to being destroyed in the impact. He headed the first expedition to the region, organised by the Soviet Academy of Sciences in 1921, which failed to get near the site of the explosion (partly because of inaccurate information about exactly where the object had fallen), but gathered a wealth of eye-witness reports. In the following few years, Kulik built up a file of information about the event, including press cuttings from 1908. He published the material in 1927, the year that he set out on the first scientific expedition to reach the site of the explosion, now pinpointed as 60° 55' 04".6 North, 101° 56' 55".6 East. The site was 70 km from the nearest habitation, a trading post at Vanavora, and could only be reached by trekking through mosquito infested taiga – bog and swampy forest crossed by countless tiny streams and rivers. They found a region of devastation that stretched as far as the eye could see, with trees felled far into the distance. In his diary, Kulik described his impressions of the scene:

> From our observation point no sign of forest can be seen, for everything has been devastated and burned, and around the edge of the dead area the young twenty-year-old forest growth has moved forward furiously, seeking sunshine and life. One has an uncanny feeling when one sees giant trees snapped across like twigs, and their tops hurled many metres to the south.

Further expeditions to the scene of this devastation followed in 1928 and 1929–30. There was then a ten-year gap before the scientists returned in 1939, when Kulik headed his fourth and final expedition to the scene. He died, while a prisoner of war, in 1942.

In 1947, the fall of another large meteorite in Siberia (more of this shortly) distracted the attention of Soviet scientists from the Tunguska site, but another full-scale expedition took place in 1958, and since 1963 there have been annual visits which now, with the end of the cold war, include scientists from outside the former Soviet Union. Their discoveries have confirmed the nature of the object that exploded over Siberia in 1908, and pinned down the size of the devastated region: forest was flattened over a butterfly-shaped area 15–35 km across. But nothing brings home the scale of the event more clearly than the eye-witness accounts gathered by Kulik in the 1920s.[4]

The only people living in the affected region were nomads. To get a picture of life in this wilderness, with nomadic tribes and trading posts, you won't be too far wrong if you imagine the age of pioneering in North America. One woman told how:

> Early in the morning when everyone was asleep in the tent it was blown up into the air, together with the occupants. When they fell back to earth, the whole family suffered slight bruises but Akulina and Ivan actually lost consciousness. When they regained consciousness they heard a great deal of noise and saw the forest blazing around them.

This is typical of many reports from the Evenki people who were on the fringes of the blast zone itself. Three eye witnesses told of the effect of the impact at Vanovara, 70 km from the epicentre of the explosion. S. B. Semenov told how:

> I was sitting in the porch of the house at the trading station of Vanovara at breakfast time and looking towards the north . . . suddenly the sky was split in two, and high above the forest the whole northern part of the sky appeared to be covered with fire. At that moment I felt a great heat as if my shirt had caught fire; this heat came from the north side. I

wanted to pull off my shirt and throw it away, but at that moment there was a bang in the sky, and a mighty crash was heard. I was thrown to the ground about three sazhen [5–6 metres] away from the porch and for a moment I lost consciousness . . . the earth trembled, and when I lay on the ground I covered my head because I was afraid that stones might hit it. At the moment when the sky opened, a hot wind, as from a cannon, blew past the huts from the north.

The second eye witness from Vanovara, Semenov's daughter Kosolapova, recalled that:

I was 19, and when the meteorite fell I was at the trading station of Vanovara. Marpha Bryukhanova and I had come to the spring for water, Marpha began to draw the water, and I was standing beside her facing north. Suddenly before me I saw the sky in the north open to the ground and fire pour out. We were terrified, but the sky closed again and immediately afterwards bangs like gun-shots were heard. We thought that stones were falling from the sky, and rushed off in terror, leaving our pail by the spring. I ran with my head down and covered, because I was afraid that stones might fall on it, and Marpha ran beside me. When we reached the house, we saw my father, Semenov, lying unconscious near the barn opposite the porch of the house. Marpha and I carried him into the hut. Whether it was hot or not when the fire appeared I don't remember. We were terribly frightened at the time. The fire was brighter than the Sun. During the bangs the earth and the huts trembled greatly,

53

and earth came sprinkling down from the roofs. The noises at first were very loud, and seemed to be right above our heads, and then they became quieter and quieter.

The third of the eye witnesses was one of Semenov's neighbours, Kosolapov:

> In June 1908, about 8 am, I was at the trading station of Vonovara, getting ready for hay-making, and I needed a nail. Not being able to find one in the hut, I went out into the yard and began to pull one out with pliers from the window frame. Suddenly, it was as if a fierce heat scorched my ears. I held them, thinking the roof was on fire, looked up and asked Semenov, who was sitting on the porch of his house, 'Did you see anything?' 'Of course I did,' he answered . . . I went straight into the hut, but no sooner was I in and about to sit down to work on the floor than a crash echoed, earth came sprinkling down from the ceilings, the oven door flew off the Russian stove on to the bed that stood opposite, and one window pane in the hut was smashed. After this, a sound like the roar of thunder was heard, getting farther and farther away towards the north.

Kosolapov's account helps to fill in the details of what occurred. He was standing by the south-facing wall of his hut, shielded from the site of the explosion to the north. So the heat he felt on his ears was a scorching produced by the trail of the fireball across the sky, not the explosion. And as his account makes clear, there was a gap between the flash of the explosion to the north, which made

Semenov want to tear the shirt off his back, and the shock wave which threw Semenov to the ground and shook the buildings. Semenov, dazed by the concussion, remembered the flash of light and the shock that threw him to the ground as being almost simultaneous, but the other two accounts make it quite clear that there was a gap between them, as there should have been – due to the difference between the speed of light and that of the seismic disturbances, moving at the speed of sound in rock.

There are many other eye-witness accounts of the Tunguska Event, both from tribespeople close to the site and many reputable people (including police officers) further afield. But this is enough to make our point. The Tunguska Event involved many of the ingredients invoked to account for the death of the dinosaurs – a devastating explosion, forest fires, a temporary reduction in the amount of heat from the Sun reaching the surface of the Earth – but on a smaller scale. Magnify all of these effects a million times, and you may begin to imagine the destruction that the Earth experienced at the end of the Cretaceous.

But imagine, also, the effect of such a disaster striking a populated part of the world in pre-historic times. Victor Clube and Bill Napier, two British astronomers, have pointed out, in their book *The Cosmic Serpent*, the similarity between the eye-witness accounts of the Tunguska Event and, for example, the Biblical account of the destruction of Sodom and Gomorrah. Fanciful? Maybe, but we can only decide that by finding out just how often events like the Tunguska explosion are likely to occur. Have they been common throughout recent times, or are they so rare that they can safely be ignored?

One of the most incautious comments about the Tunguska Event was made by Roy Gallant, in the magazine *Sky & Telescope* in June 1994. He said that 'one cannot dispute that it was responsible for the most devastating assault on our planet in the history of civilization.' One certainly can dispute this: what records would we have today of any comparable event that had occurred in the Amazon jungle, or in Siberia itself, or some other desolate region of the globe a few hundred years ago? Even leaving aside the possibility that some of the stories from ancient times of fire falling from the sky might be based on fact, not just on myth and legend – besides, where do myths and legends come from? And Gallant's comment seems particularly unwise in the light of two events that have occurred since the Tunguska impact.

We have already mentioned one of these: the meteorite that also fell in Siberia, some 5,000 km east of the Tunguska site, in 1947. Known as the Sikhote–Alin meteorite, it hit the Earth at 12.30 am (GMT) on 12 February. It is the second largest known impact (after the Tunguska event) of the twentieth century. It struck at 46° 9′ North and 134° 39′ East, only some 375 km from the port of Vladivostok. An object with a mass estimated at 100 tonnes disintegrated in the air and spread sizeable pieces of debris across the taiga in a meteorite shower concentrated in an elliptical region about 2 km long and 1 km wide. The biggest individual crater found in this region was 30 m wide and 6 m deep, and contained a two-tonne chunk of the meteorite. Altogether, investigators found 120 craters, each more than half a metre across, and recovered some 70 tonnes of meteoritic material.

Like the Tunguska event, the blast felled trees, splaying them out from the centre of the explosion. As at Tunguska 39 years earlier, eye witnesses described the fireball streaking across the sky and explosions that blew out window panes and made the ground tremble beneath their feet. Like the Tunguska object, the Sikhote–Alin meteorite blew up about 6 km above the ground; but because it was smaller and the explosion was weaker, and also because it was largely made of metal, many substantial fragments survived to reach the ground. But 'weaker' by the standards of meteorite impacts should not be taken to mean 'weak' by any human standards. Some estimates of the explosive power of the Sikhote–Alin impact put it as high as a few megatonnes, and it was certainly at least 100,000 kilotonnes, much bigger than the explosion that devastated Hiroshima in 1945.

That makes two impacts, each big enough to devastate a city, striking in roughly the same part of the world within 40 years of each other. This is already enough to make you think twice about dismissing the Tunguska Event as 'just one of those things'.

Then, in April 1972, came a genuine close encounter. Visitors to the Rocky Mountains were treated, one bright afternoon, to the sight of a spectacular fireball travelling across the sky from south to north, leaving behind it a glowing rain produced by tiny meteors, fragments from the main object, burning up in the Earth's atmosphere. The object, which was also detected by US Air Force satellites, skimmed through the Earth's atmosphere 55 km above Montana, and headed back out into space above Edmonton, in Canada. Nobody knows exactly how big it was, but from the appearance of the fireball it was caused by an object substantially bigger than the one which struck Tunguska in 1908, and the satellite data suggest a mass of about 1,000 tonnes. Two direct hits and a widely observed near miss in the space of 65 years is beginning to look like a cause for concern. Especially when you consider that similar events to the Rocky Mountain firework display of 1972 may have gone unobserved over the oceans (or above the clouds on overcast days), or unremarked even if they occurred over land in many parts of the world in the first half of this century.

And there was another near miss, even more recently, that made fewer headlines than might have been expected. On 1 February 1994, another spectacular meteor skimmed through the Earth's atmosphere. It is perhaps fortunate that this did not occur a few years earlier, during the tensions of the cold war. The large fireball produced by this object could have been mistaken for an incoming nuclear device, perhaps triggering retaliation. It was spotted by six US spy satellites, and seemed so alarming that, according to widespread but unconfirmed reports at the time, President Clinton was woken by his defence staff, uncertain whether the event was natural or man-made.

This meteor came in on a northwest to southeast path, entering the atmosphere just north of the equator, over the Pacific islands of Micronesia, north of New Guinea. It crossed the equator and exploded 20 km above the sea near the island of Tokelau, north-west of Fiji. It travelled at about 20 metres per second (72,000 km per hour), had a mass of more than a thousand tonnes, and exploded with a force of about 100 kilotonnes of TNT. Fishermen who saw it described the fireball as being nearly as bright as the Sun. It left a huge trail of smoke and dust that was still visible an hour later.

But there is one key problem in trying to assess just how real the threat of more impacts of this kind occurring in the near future is. The Earth is a living planet, covered by vegetation in many places, and eroded by wind and weather everywhere. The traces of old meteorite impacts are hard to find – even the Tunguska site is now covered by a fresh growth of trees, less than a century after the explosion. So the best place to look to get some idea of the risk of our own planet being struck by a meteorite is not at the surface of the Earth itself, but at our nearest neighbours in the inner part of the Solar System – Mercury, Venus, Mars and our own Moon. The battered faces of these lifeless bodies provide a much more accessible record of the frequency of past impacts, and show what a dangerous place the inner Solar System is to live in.

Notes

1 At one time, a wild suggestion that the explosion was caused by a fragment of antimatter striking the Earth and causing a kind of super nuclear explosion gained some publicity, but there is no trace of the radiation that would have been produced by such an event at the site, and no need to invoke such 'explanations' when a standard meteoritic impact will do the job more than adequately.
2 English translation by J. S. Romankiewicz, published by Pergamon Press in 1966.
3 And as if all that were not enough to cause confusion in the ranks of scientists engaged in investigating the Tunguska Event: by an odd coincidence, one of the British meteorologists who studied the blast waves from the explosion and wrote papers about the event in the 1930s was named Francis John Welsh Whipple; nothing to do with the American astronomer Fred Lawrence Whipple, a comet expert who also published papers about the Tunguska Event, and who we shall meet again soon.
4 History of the Bolide of 30 June 1908, in *Reports of the Academy of Sciences of the USSR*, 1927, p. 393. Most of the relevant material can also be found in Krinov's book, already referred to; Krinov was Kulik's Deputy Leader on the 1929–30 expedition.

IV

BATTERED WORLDS

To modern eyes, the battered face of the Moon is the archetypal example of an object that has been pounded by meteorites. But until the time when Galileo first looked at the Moon through a telescope, at the end of the first decade of the 17th century, the Moon was thought to be a perfectly smooth sphere, with its surface markings no more significant than a coat of paint. Galileo shattered this illusion, writing in his *Siderius Nuncius* ('Starry Messenger') that:

> The surface of the Moon is not perfectly smooth, free from inequalities and exactly spherical, as a large school of philosophers considers . . . on the contrary it is full of irregularities, uneven, full of hollows and protuberances, just like the surface of the Earth itself, which is varied everywhere by lofty mountains and deep valleys.

As telescopes improved over the centuries, it became clear that – as well as mountains and valleys, and the seemingly smooth plains which were called 'seas' (Maria in Latin) because they were thought, in the 17th century, to resemble the oceans of the Earth – the Moon is covered in craters. The craters are just beyond the limit of visibility of the unaided human eye, but even a pair of ordinary binoculars (about as powerful as Galileo's original telescope) will show up some of the larger ones, and more powerful telescopes reveal the existence of tens of thousands of them. There are more craters in the highland regions – so many, in fact, that the craters overlap one another, and there are no gaps between them – but there are also craters in the Maria, which are not seas at all, but regions that were flooded with molten rock long ago. There are about 30 times more craters in a given area of the lunar highlands (which make up 83 per cent of the lunar surface) than in the same area of Maria. Altogether, there are more than 30,000 craters each larger than 12 km across on the surface of the Moon.

When spacecraft visited the Moon, their cameras showed that every part of the lunar surface, including the interiors of large craters, is pockmarked with small craters. And most of the 'lunar seas' are themselves large circular basins, resembling huge craters. The largest obvious circular structure on the Moon, Orientale Basin, is on the far side and has never been seen from Earth; if it had been, probably nobody would ever have doubted that the craters are impact structures. Orientale Basin is a classic bullseye feature, about 700 km across (roughly a quarter the size of Texas) and surrounded by mountain rings, a spectacular sight in photographs taken by lunar orbiting spacecraft. The impact that created the basin, about 3,800 million years ago, was so powerful that it threw out lumps of rock that created secondary craters 2,000 km away.

The first strong case for an impact origin of the lunar craters was made as long ago as 1893, by the geologist

Grove Gilbert. The idea did not catch on immediately, and it was only in the 1940s that it really began to be taken seriously. Right up until the time of the Apollo missions, at the end of the 1960s and into the early 1970s, there was, however, a rival school of thought which argued that many of these craters and circular features were a result of volcanic activity. But the idea that the craters were produced by impact very much had the upper hand by the 1960s, and was proved beyond doubt by studies carried out during the Apollo missions, photographs of the Moon from close range, and analysis of rocks brought back to Earth by the Apollo astronauts – samples which included classic glassy tektites produced in the meteoritic impacts.

The Moon is scarred by impacts on all scales, from modest craters produced by the impacts of small objects to major circular structures blasted out by incoming asteroids hundreds or even thousands of metres across. The Imbrium Basin, which forms one of the 'eyesockets' of the 'Man in the Moon', has a diameter of 1,500 km and was created by the impact of an object about 40 km in diameter. Even the Maria are super-craters, flooded with lava that welled out from within the Moon in the aftermath of the impacts that formed them.

But most of the activity that has shaped the face of the Moon, and given it the familiar 'Man in the Moon' appearance, took place long ago. Analysis of Apollo specimens showed that in many cases they had lain within 10 cm of the surface of the Moon, exposed to radiation from the Sun, for at least a hundred million years. Radioactive dating of samples brought back from the Sea of Tranquillity by Apollo 11 showed that they were 3.8 billion years old – and the landing site for Apollo 11 had been carefully chosen because it was in a smooth part of the Moon, relatively free from craters. The lack of craters means that the region, like the other lunar seas, is *young*, compared with the densely cratered highland regions. If the lunar seas were as old as the lunar highlands, they would be covered in the same profusion of craters. The solidified lava from the Sea of Tranquillity may be 3.8 billion years old, but it is among the youngest rocks on the surface of the Moon.[1] And that also told astronomers, immediately, that the massive bombardment which had produced the overlapping craters of the lunar highlands must have almost ceased by the time the Maria formed.

Detailed investigations showed that there may have been several episodes of bombardment early in the history of the Moon, ending with a 'terminal lunar cataclysm' about 4 billion years ago. Since then, a much more modest, but relatively steady, rain of cosmic debris seems to have fallen on the lunar surface. Crucially, the number of craters seen in the Maria tells us how often impacts on the Moon have occurred over the past three billion years or so – how many impacts of a particular size strike each region a particular size (say, 100 km square) in every million years.

Although the events of the past few hundred million years may not have been as spectacular as what happened around 4 billion years ago, the surface of the Moon is smothered with craters formed in the past 600 Myr (that is, since the end of the Precambrian on Earth). There are about 5,000 such 'young' craters more than 5 km across, and within any circle 100 km in diameter drawn at random on the Moon's surface you will find 500 craters at least a kilometre in diameter, all younger than 600 Myr. This is one of the most important pieces of evidence in assessing how many meteorites of each size have struck the Earth over that timescale, and how many are likely to arrive in the next hundred, or thousand, or million years. Crater counting tells us that a really big impact still strikes the Moon every few million years.

The biggest impact of all, however, took place long ago, during the great bombardment. In the mid-1990s, the NASA probe Clementine compiled the most complete topographic map of the Moon yet. It revealed the presence of a feature 13 km deep and 2,500 km across on the far side of the Moon. It was probably created by the glancing impact of an object 200 km across, near the south pole of the Moon, which pushed material up to form the highland region north of this basin, dubbed the Aitken Crater.

Another recent discovery, intriguing in the light of what happened to comet Shoemaker-Levy when it went too close to Jupiter, is the presence of chains of craters on the Moon. Some of these chains are simply the result of material splashed out by a large impact falling back to the Moon. But two chains in particular have now been identified which cut right across pre-existing craters, and can only be explained by the impact of several pieces of material one after the other. The explanation is that they were the fragments of comets disrupted by the gravitational field of the Earth. One chain, called the Davy chain, is a line of 23 craters, 47 km long; the other, the Abulfeda chain, consists of 24 craters in a line 250 km long. Each of the craters in the Davy chain is a couple of kilometres across; those in the Abulfeda chain are 5–13 km in diameter. Jay Melosh, of the University of Arizona, interprets them as evidence that a sizeable comet breaks up near the Earth about once every 10,000 years. The time-scale is significant, because once you start dealing with events that take place every few thousand years, rather than at intervals of many millions of years, you have to consider the likelihood (no longer merely a possibility) that they have occurred during the time that our civilization has existed; and debris from such a cometary break-up may well spatter the Earth from time to time, not just the Moon.

But where did the Moon come from in the first place? In recent years, astronomers have come round to the view that the Moon's very existence may be a result of one of the most spectacular cosmic encounters that a planet the size of the Earth can experience and still survive.

The origin of the Moon is tied in with the origin of the other planets of the Solar System, and the origin of the Sun itself. Although there have been several hypotheses trying to account for the formation of the Solar System, the front runner is the idea that the planets formed from a disc of dusty material around the young Sun. The astrophysicists cannot yet explain exactly how such a disc of interstellar debris can produce planets, but their confidence in the idea has been boosted by studies of objects such as Beta Pictoris – a young star embedded in a disc of dust that has actually been photographed, and which extends out on either side of the star across a region several times larger than our own Solar System.

In such a disc, tiny grains of dust will collide and stick together. This process probably began in the collapsing cloud of gas and dust before the star at the centre had got hot enough to light up. By the time the Sun had formed, it was surrounded by clumps of matter perhaps a few millimetres across, settling into a disc. The primordial pebbles gradually clumped together to form objects like asteroids, perhaps a kilometre or so in diameter. At that point the gravity of the individual objects became important, tugging them together into swarms where they bumped into one another, interacting and colliding, sticking together in larger and larger lumps. The largest lumps, with the strongest gravitational pull, would soon attract the smaller lumps, until they formed planets and moons.

One important feature of this understanding of the formation of the planets is that there must have been plenty of cosmic debris left over from the planet-building process. The pattern that has emerged, in the case of our own Solar System, is one in which there are four relatively small, rocky planets close in to the Sun (Mercury, Venus, Earth and Mars), and four relatively large, gaseous planets further out from the Sun (Jupiter, Saturn, Uranus and Neptune). Between the 'terrestrial' planets and the 'gas giants' there is a band of cosmic rubble, known as the asteroid belt, made up of leftover material from the formation of the planets. This debris could not form into one lump because it is disturbed by the gravitational pull of Jupiter, by far the largest planet in the Solar System. A ninth object, Pluto, is generally regarded as a planet, even though it is a small object in a very peculiar orbit which sometimes takes it further from the Sun than Neptune and sometimes (including during the 1980s and 1990s) closer to the Sun than Neptune.

But even after the four major 'lumps' that were to become the inner planets had formed in the disc around the Sun, there were still many lesser objects around, being swept up by the proto-planets as they orbited through the disc. The picture that emerges is one in which all of the inner planets should have been intensely bombarded early in their lives; and in which the 'terminal lunar cataclysm' (which perhaps might better be called the 'initial lunar cataclysm'!) can be seen as the natural end-phase of a process of planet building that began about 4.5 billion years ago, when the Sun formed, and ended about 4 billion years ago. And somewhere in that half-billion year interval, it now seems that the Moon was ripped out of the Earth itself, by the impact of an object about as big as Mars.

The first person to suggest anything like this origin for the Moon was George Darwin (one of the sons of Charles Darwin), who proposed that the Moon was a piece that had split off from the Earth early in its development, and been flung away into space. This simple fission effect won't work – in order to fling a chunk the size of the Moon away the early Earth would have had to be spinning impossibly fast. But things are different if material from the Earth is knocked away into space in a cataclysmic collision. The impact idea was first put forward in 1946, by two American geologists, Ralph Baldwin and Don Wilhelms. Their suggestion was completely ignored at the time, and the modern version of the idea developed independently from work by William Hartmann and Donald Davis in 1975.

You shouldn't think of this process as like chipping a piece of stone from a rock. The colloquial name astronomers use for the theory – which is backed up by detailed computer simulations of what goes on in such a collision – is the 'Big Splash'. The impact of a Mars-sized object with the young Earth would have produced so much heat that the proto-planet was left with a global ocean of molten rock about a thousand kilometres deep, while the incoming object would have been completely disrupted in the collision. Any dense core of metallic material that it contained would have been captured by the Earth, subsiding through the molten surface layers to merge with the Earth's own iron core. Meanwhile, the lighter parts of the incoming planetoid, together with parts of the outer crust of the young Earth, glowing red-hot chunks and molten globs of rock, would have been flung off into space.

The hot debris from the collision would form a ring around the Earth, from which all light substances, such as water, would have evaporated away. Then, the pieces would have cooled and gradually coalesced to form the Moon, echoing the way in which the planets formed from rings of material around the Sun. As a byproduct, the impact of the collision would have made the Earth spin faster on its axis, giving us a relatively rapid succession of day and night. (Venus, Earth's near twin in size, has no moon and spins much more slowly on its axis, with each day lasting a year – circumstantial evidence in support of the theory.) By hitting the Earth off-centre, the impact probably also caused the tilt in its axis which is responsible for the variation of the seasons over the year. So our lives today, four billion years later, are very much governed by the consequences of the impact with the Earth in which the Moon was created.

This model of how the Moon formed is now very well established, both by computer studies and by investigations into the composition of the Moon itself. It does not seem to contain even a trace of water or other volatile materials, and also is the only rocky world in the inner Solar System without an iron core. Moreover, it is the only large moon orbiting any of the inner planets – although only a quarter the size of the Earth, the Moon is so large compared to other moons (in proportion to their 'parent' planets) that the Earth–Moon system should, strictly speaking, be regarded as a double planet. A clinching piece of evidence in support of the Big Splash model, in the eyes of many astronomers, is that the Moon orbits the Earth at a pronounced angle to the equator of our planet, suggesting that it did not form alongside the Earth out of the same primordial swarm of cosmic debris.

The object which most resembles the Moon in the inner part of our Solar System is the planet Mercury, 38 per cent of the size of the Earth and the closest planet to the Sun. The Moon-like appearance of Mercury came as a complete surprise to astronomers when pictures from Mercury were first beamed back to Earth by the spaceprobe Mariner 10 in the mid-1970s. With the hindsight of the standard model of the Solar System developed since then it is less surprising. Mariner 10 provided pictures of the face of Mercury similar in quality to images of the surface of the Moon provided by a high-powered telescope. It revealed a battered planet scarred with craters, in places resembling the battered highland regions of the Moon, although on Mercury there are some gaps between the craters. All the evidence is that the craters were produced at the same time that the Moon was receiving its battering, about 4 billion years ago, with a lesser but steady rain of impacts since then.

The most spectacular impact feature on Mercury is the Caloris Basin, about 1,300 km in diameter, which is surrounded by a ring of mountains, 2 km high, created in the impact. The basin is big enough to contain the British Isles within its ring of mountains. But the most spectacular signs of the impact are on the other side of the planet, diametrically opposite the Caloris Basin, where there is a region of jumbled terrain created by shock waves which travelled right round the planet and threw up jumbled hills. Imagine that hills in Britain were thrown up by shock waves from a meteorite striking New Zealand, and you get some idea of the relative importance of the impact that created the Caloris Basin for Mercury.

All of this cratering is preserved for us to see today because, like the Moon, Mercury has no atmosphere, so there is no erosion to wipe away the cratering record. One difference between the Moon and Mercury is that Mercury has a stronger gravitational pull, so that material ejected by impacts cannot travel so far before falling back to the ground. Another is that Mercury is (or has been) more active internally than the Moon, starting out with a hot core and cooling off as time passed, so that the planet has shrunk slightly, wrinkling its surface a little as it did so. Internal activity may also explain the gaps between the craters on Mercury: possibly the original surface was covered by craters, like the highland regions of the Moon, before some internal upheaval leaked lava onto the surface, smoothing out some of the craters and providing a fresh target for later missiles.

But the most important message gleaned from the study of photographs of the battered face of Mercury is simply that whatever happened to the Moon when it was young was not unique. It was not a particular consequence, for example, of the fact that it was indeed formed in orbit around the Earth, where, it might have been argued, the cosmic battering was simply a result of gathering up the last pieces of the debris out of which the Moon formed after the Big Splash. That is clearly not the case. Without doubt, the rain of cosmic debris extended right across the inner Solar System, affecting all of the terrestrial planets.

Just what effects that rain of debris had on the other three terrestrial planets we cannot be sure, because unlike Mercury and the Moon they are each objects with atmosphere that have worn away traces of any impact features over the eons. Venus, although almost exactly the same size as the Earth, has a dense, hot atmosphere mainly consisting of carbon dioxide, laced with clouds of sulphuric acid droplets from which sulphuric 'acid rain' falls. Spaceprobes that have landed on Venus survived only briefly before being simultaneously crushed, corroded and cooked. But they managed to send back data showing the surface to be the nearest thing in the Solar System to the Biblical vision of Hell – a searingly hot desert under an eternally overcast sky, completely lifeless, with a surface atmospheric pressure 90 times that of the Earth and a surface temperature of over 450 degrees Celsius.

The information we have about the surface topography of Venus comes mainly from the spaceprobe Magellan, which went into orbit around the planet at the beginning of the 1990s, and mapped almost the entire surface by radar. The radar maps show that Venus is a highly active planet, covered by volcanoes and lava flows, with a cracked, tortured surface. Over all of the surface there is also a uniformly random pattern of impact craters. But there are far fewer craters than there would be if, like Mercury and the Moon, Venus had retained its ancient surface layer. From comparison with the number of craters on Mercury and the Moon, astronomers infer that the entire surface of Venus was wiped clean in some great cataclysm about 600 million years ago. Geologists describe this event as having 'resurfaced' the planet with lava from its interior, as great blocks of the surface cracked and subsided; in all probability such events have occurred several times in the 4 billion years or so since Venus formed.

In some ways, from our point of view, this recent, more modest cratering of Venus is even more disturbing than the more comprehensively battered terrains of the two airless worlds. The surface of Venus shows clearly that objects, large enough to survive the passage through the planet's very dense atmosphere and produce scars on its surface tens of kilometres across, have rained down steadily over a timescale almost exactly corresponding to the length of the detailed fossil record on Earth – the time since the Precambrian. Many of the craters seem to have been formed by 'airblasts', when meteoritic objects exploded in the atmosphere of Venus, just like the Tunguska Event but on a much larger scale. There are hundreds of craters 'visible' to the radar eyes of probes like Magellan – 400 air-blast craters alone have already been identified, and there may be as many again which have not yet been counted, plus all the visible impact craters. This is a modest number compared to the overall cratering of the Moon and Mercury, but it means impacts on this scale occurred at least every million years or so for the past 600 million years. There seems no way to escape the conclusion that Earth has been subjected to a similar bombardment from space over the same sort of timescale.

This kind of impact doesn't just scar the surface of a planet; as the example of Mars shows, it can drastically alter its atmosphere as well. Mars is only 53 per cent of the size of the Earth, and orbits further out from the Sun than we do. It now has only a thin carbon dioxide atmosphere, and the surface temperature is well below freezing. But when spaceprobes visited the planet and went into orbit around it in the 1970s, as well as finding the ubiquitous craters that scar all of the inner planets, they photographed features that could only have been produced by running water (or, just possibly, some other flowing liquid; but there are no other contenders).

Erosion on Mars has smoothed out the smaller, older craters, but counts of craters on Mars show that the smooth plains of the planet are roughly the same age as the Maria of the Moon, so it has not experienced the kind of cataclysmic upheaval that turned Venus inside out 600 million years ago. Many Martian craters are, allowing for the erosion by the atmosphere, very much like lunar features. Argyle Basin, for example, is a feature 600 km across, blasted out of the Martian surface by a major impact and floored with lava released by the impact. Also like the Moon, Mars possesses heavily cratered highland regions. And like Venus, it possesses some impressive volcanoes, including Olympus Mons, a feature 500 km across and towering 23 km above the surrounding plains. The largest volcano on Earth (which is itself almost twice as big as Mars), Mauna Loa in Hawaii, is only 200 km across and rises just 9 km above the sea floor.

Like the Earth, Mars once had running water. Even some of the craters show this. Instead of being sharply edged like the craters of the Moon and Mercury, the walls of these craters slump into their surroundings, as if the impacting object had plopped onto a wet sludge of material. At the time this happened, the water may well have been frozen below the surface of Mars, thawed in the impact to make the sludge, and then refroze afterwards. But still, there must be water there to do the trick. Not only that, the water must once have flown fairly freely across the surface. One winding valley, 575 km long, could only have been carved by a river, while the walls of the great canyons that split the surface of Mars are also marked by erosion caused by running water. At other sites 'streamlines' show how material has been deposited by water flowing around obstructions.

In order for water to flow on Mars, its temperature must have been higher long ago. Yet astronomers calculate that, if anything, the Sun was cooler when it was younger than it is today. The only way the temperature at the surface of Mars could have been kept above freezing would have been if the planet had had a thicker atmosphere, in which the greenhouse effect (which works by trapping heat from the Sun and preventing it escaping easily back out into space) could have got to work. Where could such a thick atmosphere have come from? And where has it gone?

There are two schools of thought about the origins of the atmospheres of Venus, Earth and Mars. One holds that volcanic activity is all you need, outgassing material from inside the planet. The other suggests that atmospheres – and in the case of the Earth, oceans – were built up soon after the planets formed, by cometary impacts. By studying the cratering record on the Moon, Christopher Chyba, then of Cornell University, calculated in 1987 that a total of more than 10^{19} tonnes of material from space should have arrived on Earth during the intense bombardment that occurred between 4.5 and 3.8 billion years ago. This incomprehensibly large number – a 1 followed by nineteen noughts, or 10 billion billion tonnes – is so big that if only 20 per cent of the incoming material were in the form of comets, and only 50 per cent of the mass of each comet was water, then the overall effect would be to supply 80 per cent of the present oceanic mass, with the rest to be topped up by volcanic out-gassing.

It is even easier to make atmospheres in this way since they are much less massive than oceans. So when Mars was young, it probably had a much thicker atmosphere than it has now, provided by cometary impacts which also contributed at least some of the water which ran across its surface to make the distinctive features still visible today. But the surface gravity on Mars is much weaker than that on Earth, because Mars is a smaller planet, and so it has trouble holding on to any atmosphere it started out with. For a long time, astronomers thought that Mars might simply have lost the bulk of its atmosphere slowly, as molecules leaked away into space over the eons. In that scenario, it could have remained watery for some considerable time. But in 1989 Jay Melosh and Ann Vickery pointed out that large meteorite impacts would give a very efficient push to these molecules, stripping away the atmosphere in a series of violent events.

Because the gravity on Mars is so weak, it is easy for the expanding cloud of vapour from a major impact to blast all of the atmosphere in its vicinity out into space. Just how big an impact you need to do the trick depends both on the size of the planet involved and the density of its atmosphere (thicker atmospheres are, of course, harder to push aside). So whereas on Venus you would need an impacting object 70 km across to blow a significant amount of atmosphere away into space, on Earth you need one 13 km across, and on Mars all you need is a cosmic rock 3 km in diameter.

Working backwards from the present state of Mars to calculate how it started out (in effect, taking away big impact craters one by one and adding back the atmosphere which would have been lost in each impact), Melosh and Vickery calculated that Mars probably had an original atmosphere with about the same surface pressure as that of the Earth today, and a correspondingly high surface temperature, above the melting point of ice. But this more clement Mars was already disappearing, on the new picture, 4 billion years ago, its atmosphere being torn away in the Martian equivalent of the terminal lunar cataclysm.

The low surface gravity on Mars also means that it is relatively easy for large impacts to blast fragments of Martian surface rock out into space, escaping from the planet's gravitational grip. Astonishingly, some of these Martian fragments have arrived on Earth, falling as meteorites. Originally, just eight of these Martian meteorites were known (a few more have been found since). They were identified as coming from Mars in the early 1980s, partly because standard dating techniques showed them to be much younger than 'ordinary' meteorites, and partly because analysis of gas bubbles trapped in the fragments of rock showed the same composition as gases in the atmosphere of Mars, analysed by the Viking landers.

It was only after these findings that researchers realised that some other small meteorites are actually samples of Moon rock, blasted out in a similar way – the composition of these meteorites matches samples of Moon rock brought back by Apollo astronauts. So these lunar meteorites are made of material which probably once formed part of the crust of the Earth, before being blasted into space by a giant meteorite impact, becoming part of the Moon, and then being blasted back out into space in a lesser impact and falling back to Earth. The puzzle was how fragments of rock could have been blown out intact from the Moon and Mars by impacts, without being melted and splashed, like tektites (incidentally, it is quite possible that some tektites found on Earth have come from the Moon or Mars).

Melosh and Vickery showed how the shock wave from an impact can actually cause the surface of the planet to bounce, flicking boulders up to 15 metres across off the surface of Mars and into space. Their calculations suggest that the best way to achieve this would be a single impact producing a crater 200 km in diameter about 200 Myr ago (the age is worked out from the extent to which the rocks have been altered by exposure to radiation during their sojourn in space). Unfortunately, there is no crater the right size, the right age and in the right part of Mars to match the composition of the eight meteorites. Another possibility, proposed by John O'Keefe and Thomas Ahrens, of CalTech, is that a shallow-angle impact could produce a jet of hot vapour that would blow boulders out into space like the jet from a high-pressure fire hose. That would only need a crater 50 km across, and there are plenty of candidates.

Whatever the exact mechanism involved, there is now no doubt that impacts on Mars have been severe enough, relatively recently in terms of geological time, to send pieces of the red planet across space to Earth. Some of these pieces of Martian rock contain carbonates, which can only have formed from water in which carbon dioxide was dissolved; one actually contained a few milligrams of water. And there are even traces of complex organic molecules, built up from carbon dioxide, water and ammonia. All this is strong supporting evidence for the idea of a warm early Mars with a thick atmosphere that has been stripped away in cosmic impacts. But, of course, if impacts can strip away atmosphere from Mars, then suitably larger impacts – comparable in size to the terminal Cretaceous event – can strip away significant amounts of atmosphere from the Earth. Our planet, too, may have once had a thicker atmosphere which has been depleted over the eons.

This depletion may be a good thing. We mentioned that the Sun was cooler when it was young, and in those days a thicker atmosphere with a stronger greenhouse effect would have helped to keep the Earth warm. But as the Sun has become hotter, a strong greenhouse effect could have raised temperatures on Earth high enough to make life difficult. Impacts from space are not all bad news, it seems, and may have helped to thin out the atmosphere of our planet enough to keep it cool and comfortable as the Sun has warmed up. It is even possible that it was only the impact which created the Moon that prevented the Earth from having a super-thick atmosphere like that of Venus, under which a runaway greenhouse effect would have raised temperatures so high that all the water evaporated and life never got a grip.

What else can the new understanding of the battering that the inner worlds of the Solar System have received tell us about events on Earth? To start with, it is even possible that life on Earth was itself brought into being by those impacts. Whatever its origins, life could not possibly have emerged on the surface of our planet during the primordial bombardment. There are 35 impact features on the Moon each more than 300 km in diameter. Even if only half of them were formed during the primordial bombardment, that would mean that the Earth, which is a much bigger target than the Moon, was struck by more than 300 comparable impacts during the first 500 million years or so of its existence. The biggest of these events would have vapourised all the water on the surface of the Earth, killing any life that existed at that time. Christopher Chyba is among theorists who have suggested that life started to evolve several times when the Earth was young, and that there were a handful of 'sterilization events' before conditions settled down to the point where life could continue to evolve. This makes it all the more remarkable that the oldest traces of life on Earth are found in rocks 3.6 billion years old, laid down only about 200 million years after the end of the terminal lunar cataclysm. How could life have got going so quickly?

The earliest terrestrial fossils coincide almost exactly in time with the end of the period of heavy bombardment from space. This suggests to some researchers that the bombardment itself may have had something to do with the origin of life. In the most extreme version of this scenario, promoted strongly by the British astronomer Fred Hoyle, life evolved in clouds of gas and dust in space, and comets impacting with the Earth brought not only water but living cells to the young planet. Even without going to that extreme, however, it is now clear that comets contain exactly the ingredients needed to give life the opportunity to evolve. These compounds are revealed by spectroscopic analysis of the light from comets, and have been found in meteoritic debris on Earth (as we shall see, many meteorites are in fact pieces of comet). Compounds rich in the elements carbon and nitrogen are present in cometary material, as well as water, which is itself made up of hydrogen and oxygen. These are among the most common elements in the Universe, and are collectively referred to as CHON; it is surely no coincidence that they are also the key elements of life. And comets don't just contain these elements in a raw form. They exist in comets already packaged in the form of complex organic molecules,[2] including amino acids, which are themselves the basic units from which the proteins in all living things, including your body, are made.

Of course, such complex original material would be destroyed in a major impact like the terminal Cretaceous event. But it seems likely that some of the cometary debris, including dust from comet tails, would have fallen more gently down through the atmosphere of the Earth, a gentle rain rich in organic material. Even today, interplanetary dust particles gathered up by the Earth contribute about 300 tonnes of fresh organic material to the surface of our planet each year, while meteorites contribute about 10 kg a year. In a million years, that would add up to a lot of chemically complex stuff. Oceans of warm water, plenty of sunlight to stimulate chemical reactions, and a steady supply of organic material such as amino acids makes up the perfect recipe for starting life itself down the evolutionary road. Given those starting conditions, it would have been astonishing if life had not got a grip on Earth within a couple of hundred million years of the end of the primordial bombardment.

More speculative ideas about cosmic impacts include the suggestion that it was the impact of an asteroid about 20 km across that cracked the Earth's crust 250 million years ago, setting the pieces moving in the latest phase of continental drift, and also causing the mass extinctions at the end of the Paleozoic era. Enthusiasts for this idea have identified a possible site for the impact, in the form of a crater 300 km across in the sea floor near the Falkland Islands.

Overall, from comparison with the cratering record on Mars, Mercury and the Moon, it is possible (after allowing for their different sizes) to work out how often impacts of different sizes have occurred on Earth in recent geological times. An object bigger than 1 km in diameter strikes Mercury once every 10 million years, Mars five times every 10 million years, and the Earth 60 times every 10 million years. Such an object is big enough to cause a crater about 20 km across on Earth, and one hits the Earth every couple of hundred thousand years. An object big enough to cause a crater 30 km across strikes roughly once every million years; one big enough to make a crater 50 km across strikes every 12 million years; and one big enough to blast out a crater 100 km across strikes every 50 million years or so. And, as we have already mentioned, events producing craters on the scale of the Chicxlub feature occur every 100 million years or so.

'New' impact sites on Earth are still being discovered – for example, it was only in 1994 that Wylie Poag, of the US Geological Survey, discovered that Chesapeake Bay, site of one of the most important early English settlements in the New World, was shaped by such an impact. The crater is one of the ten largest known on Earth, about 85 km across and produced by an impact that occurred 35 Myr ago, strewing tektites across the south-eastern part of North America. But it had remained concealed by the sea itself and sediments washed down by the rivers until revealed by seismic soundings (some of them, as in the case of Chicxlub, carried out by a company searching for oil and gas).

The list goes on, but there is no need to cite more examples here. If you are not convinced by now that we live in a dangerous cosmic neighbourhood, you never will be. But although our focus of interest must be on the immediate neighbourhood of the Earth, in the inner part of the Solar System, we cannot leave our overview of the battered worlds without at least a brief mention of what has happened in the outer part of the Solar System.

As the events of 1994 showed, when objects such as Shoemaker-Levy strike the gas giant planets, they do not leave permanent scars – certainly nothing that remains for hundreds or thousands of millions of years. But the outer planets are attended by families of moons, worlds in their own right, many of which do bear the scars of meteorite impacts. Studies of these battered moons reveal more details both about the events that shaped our Solar System, and about the way planetary systems in general are formed and evolve.

Before they had close-up views of the moons of the outer planets, astronomers expected them to be inert worlds, like our own Moon. But when the two Voyager spacecraft flew past these moons and sent back pictures of them at the end of the 1970s and in the 1980s, they revealed scenes of activity, past and present, like nothing in the inner part of the Solar System.

Jupiter, the largest planet in the Solar System, is so big that it could swallow up 1,300 objects the size of the Earth, and it has at least 16 moons, spaced in four groups of four. Closest to the planet there are four small moons; further out, four giant moons, so big that if they orbited the Sun instead of Jupiter we would be happy to call them planets (each of them is bigger than Pluto); further out still are four more small moons; and far out on the fringes of the jovian system there are four tiny moons that are probably captured asteroids.

The four large moons – often called the 'Galilean' satellites, because they were discovered by Galileo with the aid of his telescope – are the most interesting ones. Callisto, the furthest from Jupiter, is about the same size as Mercury, and like Mercury it has a heavily cratered surface. These craters cover every part of the moon's surface, like the craters in the highland regions of our own Moon, and are thought to represent ancient terrain, battered in the bombardment in which the jovian system formed. One bullseye feature on Callisto, known as Valhalla, is 2,600 km across – on a moon which is itself only 4,800 km in diameter. Next in towards Jupiter lies Ganymede, 5,262 km in diameter, the biggest moon in the Solar System and almost as large as Mars. Parts of the surface of Ganymede are as heavily cratered as the surface of Callisto, but parts are less so, lighter in colour and marked with strange ridges and grooves. It looks as if the kind of tectonic activity (continental drift) that has shaped the surface of the Earth began on Ganymede, but petered out.

While Callisto is the most heavily cratered object in the Solar System, the third Galilean satellite, Europa, is the smoothest world we know. It looks like a snooker ball the size of our Moon. Close-up images show that it is completely covered by a relatively fresh layer of ice, marked by cracks and long, low ridges. The ice must have emerged as water from inside Europa only a few million years ago, probably heated by tidal effects in the strong gravitational field of Jupiter, warming the interior of the moon. The covering of ice has smoothed out any primordial craters. Apart from its extent and coldness, this reworked surface is more like Venus than any other object we know. Io, even closer to Jupiter, shows just how powerful those tidal forces can be. Astronomers were astonished when Voyager images showed active volcanoes on Io, spewing sulphurous material up from its surface. These are now understood to be a result of tidal forces squeezing and stretching Io's interior, heating it to the point where pockets of sulphur boil up and burst through the surface, covering it with layers of red, orange and yellow sulphur. Some of these sulphurous 'lava flows' stretch 200 km from the volcanoes where they have erupted. All this on an object only about the same size as our own Moon.

Saturn, the next planet out from the Sun, is smaller than Jupiter, but still the second biggest planet in our Solar System. And in some ways it can outdo even Jupiter. Saturn has the most spectacular ring system in the Solar System, it has at least 18 moons, and one of those moons, Titan, even has a healthy atmosphere and could conceivably be a home for life. With a diameter of 5,150 km, Titan is almost as big as Ganymede, and the thick nitrogen atmosphere is laced with methane, ammonia and orange clouds which contain complex organic molecules produced by lightning activity in the atmosphere. Unfortunately, though, the impenetrable cloud layer means that as yet we know nothing about the extent of cratering on the surface of Titan. Maybe one day a Titan orbiting probe equipped with radar will be able to map the surface in the way Venus was mapped by Magellan.

But there is plenty of evidence of cratering on the other moons of Saturn. Two of them, Mimas (390 km across) and Tethys (1,050 km across) are each marked not only by extensive small-scale cratering but by individual craters which cover a third of their respective worlds. One pair of irregular moons which orbit together (Janus and Epimethus, each a couple of hundred kilometres across) are clearly pieces of one original moon which has been smashed in two by a cosmic impact. Even this, though, is nothing compared with what has happened to Uranus, the next planet out from the Sun beyond Saturn.

Like Jupiter and Saturn, Uranus lies at the centre of its own miniature 'Solar System' of moons (at least 15 of them), travelling about it in their orbits. But unlike the orbits of all the other moons in the Solar System, the moons of Uranus do not orbit in the same plane as the one in which the planets orbit the Sun. The whole Uranian system is tipped on its side. As the planet moves round the Sun, there are times (for example, in 1986) when one of the poles of Uranus points towards the Sun. At those times, the view from Earth shows the moons moving around Uranus in circles like a bullseye target. A quarter of an orbit (21 years) later, Uranus is sideways on to the Sun, and we see the moons moving up and down as they trace circular orbits edge on to our point of view, vertical to the plane of the Earth's orbit around the Sun. After another 21 years, the opposite pole of Uranus is pointing towards the Sun, and so on.

How did Uranus get to be such an odd one out among the planets? The best explanation is that the planet was struck by an object as massive as the Earth, which knocked it on its side and completely disrupted its system of moons, which then re-formed in the orbits where we see them now. This history is attested by the bizarre nature of the moons of Uranus, especially Miranda. Only 480 km across, Miranda looks like a heap of cosmic junk that has been smashed apart and roughly re-assembled, put together in higgledy-piggled fashion. The pieces don't fit together anywhere very smoothly.

One feature, a cliff dubbed Verona Rupes, falls sheer for a height as great as the height of Mount Everest above the lowest point of the ocean floor of the Earth. And this on a world less than one twentieth the size of the Earth. Astronomers believe that Miranda is just what it seems to be – a heap of rubble pulled together by gravity out of the debris of one or more moons shattered in the cataclysm that tipped Uranus on its side, which has not yet had time to settle into a smoothly rounded world. Other moons of Uranus are also cratered and scarred, though none as spectacularly as Miranda.

After Uranus and Miranda, anything else must be an anticlimax. Neptune, the giant planet most distant from the Sun, is indeed dull by comparison. It has a family of at least eight moons, but cannot offer anything as spectacular as the shattered worlds of Uranus. The largest of its moons, Triton, has a diameter of 2,700 km, and is covered with ice that has relatively recently frozen over the surface, which is marked by a few seemingly fresh impact craters.

Although Neptune and its entourage are relatively dull, they help to fill out the overall picture of how planets and moons form. The four inner planets of the Solar System were bombarded, after they formed, with debris orbiting around the Sun with those planets; while in each of the giant planets' systems of moons, analysis of the cratering patterns shows that the cratering was produced by pieces of debris in orbit around the respective giant planet. Each of the four giant systems formed independently, and in much the same way that the Solar System itself formed. This is important for two reasons. First, it means that when more spaceprobes visit the outer worlds (and Galileo was due to arrive at Jupiter in 1995, just when we were writing) astronomers will have four 'case studies' which they can use to work out details about how smaller objects (in this case, moons) form around larger objects (in this case, giant planets). Their investigations should help them to develop a better understanding of how the terrestrial planets, including the Earth itself, formed around the Sun. Secondly, from these five examples (the Sun and its family of planets; the four giant planets and their families of moons) it seems that when large objects form in space they are always accompanied by a retinue of smaller objects. This makes it very likely that there are many other planetary systems in the Universe, where planets have formed alongside stars.

Intriguing though the implications are, following them up would go beyond the scope of the present book. But while the stars themselves are beyond our present scope, we have to go halfway to the stars to answer the next obvious question about the objects which continue to batter the terrestrial planets today. The primordial bombardment of the terrestrial planets ended nearly 4 billion years ago, and there is no trace of large numbers of prospective meteorites in orbit around the Sun in our part of the Solar System today. Yet ever since the original bombardment ended, as the recent cratering of the surface of Venus shows particularly clearly, there has been a steady rain of objects coming into the inner Solar System and smashing into the terrestrial planets. What are they, and where do they come from?

Notes

1 To put this in some sort of perspective, remember that there are so few fossils from more than 600 million years ago on Earth that everything before that time is lumped into one era, the Precambrian. The Moon had been more or less as we see it now for 3.2 billion years before the fossil record of life on Earth had really begun.

2 An organic molecule is one which is based on carbon and has complex chemistry; such molecules are useful for life, although not only produced by life.

V

HALFWAY TO THE STARS

There are two kinds of large cosmic objects that strike the Earth – asteroids and comets. Asteroids were discovered at the beginning of the 19th century, when astronomers were intrigued by the apparent gap in the spacing of the planets between Mars and Jupiter.

The gap shows up clearly in a simple mathematical expression which relates the distances of the planets from the Sun. As is often the case in science, the law does not carry the right name – it was first set out by the German astronomer Johann Titius in 1766, but was popularised by his compatriot Johann Bode in the 1770s.

In the form presented by Bode, the key sequence of numbers is 0, 3, 6, 12, 24 and so on, with each number after the '3' twice the preceding number. Then, you add 4 to each number. If the distance of Earth (the third planet) from the Sun is set at 10 units, then the distances of Mercury, Venus and Mars are all at the equivalent distances given by the series: 4, 7, 10, 16. There is a gap at 28, with Jupiter at 52 and Saturn at 100 units from the Sun, as 'predicted' by the law.

The discovery of Uranus in 1781, close to 196 units from the Sun, was regarded as a confirmation of the accuracy of

Bode's law, and in the early 19th century the asteroids were discovered in a belt between Mars and Jupiter, corresponding to the 'missing' planet.

Although Neptune and Pluto, discovered later, do not have orbits which match the predictions of Bode's 'law', there has recently been a revival of interest in the possibility that some simple rule of thumb of this kind must apply to the spacing of the orbits of planets in any planetary system, and to the spacing of moons in systems like those around Jupiter and Saturn.

The discovery of three planets orbiting a pulsar known as PSR B1257+12 has provided a boost for these ideas by revealing a system with properties that almost exactly match those of the innermost Solar System, made up of Mercury, Venus and the Earth. The similarities are so striking that it seems there may be a law of nature which ensures that planets always form in certain orbits and have certain sizes.

PSR B1257+12 is a rapidly spinning neutron star, containing slightly more matter than our Sun packed into a sphere only about 10 km across. As the star spins, it flicks a beam of radio noise around, like the beam of a lighthouse, producing regularly spaced pulses of radio noise detectable on Earth. It can only have been produced in a supernova explosion, long ago, which would have disrupted any planetary system the star possessed at the time. So the present planets associated with the pulsar are thought to have formed from the debris of a companion star disrupted by the pulsar.

The three planets cannot be seen directly, but are revealed by the way in which they change the period of the pulsar's pulses as they orbit around it. There is enough information revealed in the changing pulses to show that the three planets have masses roughly equal to 2.8 times that of the Earth, 3.4 times that of the Earth, and 1.5 per cent of that of the Earth. And they are spaced, respectively, at distances from the pulsar equivalent to 47 per cent of the distance from the Earth to the Sun, 36 per cent of the Sun-Earth distance, and 19 per cent of the Sun-Earth distance.

Tsevi Mazeh and Itzhak Goldman, of Tel Aviv University, have pointed out that the ratio of these distances, 1:0.77:0.4, is extremely close to the ratio of the distances of the Earth, Venus and Mercury from the Sun, which is 1:0.72:0.39. And the masses of the three inner planets in the Solar system are 1 Earth mass, 82 per cent of this mass (Venus) and 5.5 per cent (Mercury). In each case, two outer planets with roughly the same mass have an inner companion with a much smaller mass.

So Bode's law also works for the planets of pulsar PSR B1257+12, and the indications are that there is a universal mechanism for the formation of planets around stars. If it works for systems as diverse as a pulsar and our Sun, the chances are that it works for all stars, and that 'Solar' Systems very like our own may be the rule, rather than the exception, among the stars of the Milky Way.

But that is by the way. When the Sicilian Giuseppe Piazzi discovered a new world on the night of 1 January 1801, one of the reasons astronomers were excited by the discovery was that the object, given the name Ceres after the patron goddess of Sicily, orbits exactly where Bode's law says there ought to be a planet, orbiting the Sun once every 4.6 years between the orbits of Mars and Jupiter. Ceres turned out to be far too small to be considered a planet, with a diameter of just 942 km. But in March 1802 the German Heinrich Olbers found another of these 'minor planets', Pallas, travelling around the Sun at almost the same distance as Ceres (but in its own orbit, moving independently of Ceres). Two more minor planets, Juno and Vesta, were found in 1804 and 1807, respectively, and a fifth, Astraea, in 1845.

The first of the asteroids to be discovered were, of course, the largest and easiest to find (Ceres is the largest of them all); but by 1890 no less than 300 objects orbiting in the gap between Mars and Jupiter had been discovered, and now more than 2,500 residents of what is now known as the asteroid belt have been identified and catalogued. They orbit between 2.2 and 3.3 astronomical units (AU) from the Sun (1 AU is the distance of the Earth from the Sun). At these distances, they take between three and six years to orbit the Sun once.

Hundreds more have been observed from time to time, but never long enough for their orbits to be calculated accurately, and it is estimated that there may be as many as half a million of these objects bright enough to appear in photographs taken with the famous 200-inch telescope at Mount Palomar, if anyone had the time and inclination to make the effort of searching for them, and a billion more objects a few hundred metres across. But this still means that, on average, millions of kilometres of empty space separate each asteroid from its nearest neighbour. Out of all these objects, only five have diameters greater than 300 km, some 250 have diameters bigger than 100 km, and most of the catalogued asteroids have diameters of about 1 km.

At one time, it was thought that the asteroids might be the debris from a former planet that had exploded. But nobody ever thought of a good reason why a planet should explode, and in any case the total mass of all the objects in the asteroid belt put together is only about 27 per cent of the mass of the Moon, or 0.3 per cent of the mass of the Earth. Today, astronomers explain the asteroid belt not as the debris of an exploded planet, but as the fragments of a planet that failed to form – and they think they know why it failed to form.

David Hughes, of the University of Sheffield, has calculated that in the swirling disc of material from which the planets formed, in the region now occupied by the asteroid belt, there was originally enough primordial dust to make four rocky planets the size of the Earth. A comparable ring around the Sun at the distance of the Earth's orbit would have contained less material, because the ring would have had a smaller circumference. So why are there only fragments of protoplanetary material in the asteroid belt today, not a single planet with four times the mass of the Earth?

The answer seems to be that that part of the primordial disc was too close to Jupiter. The British astronomer Fred Hoyle spelled out the details of the development of the asteroid belt in his book *The Cosmogony of the Solar System*. At first, the particles in the asteroid belt travelled around the Sun in circular orbits, like runners running around a circular track, each in their own lane. Occasionally, particles would bump together fairly gently, nudging into the 'lane' next door and sticking to one another. But as large objects started to form by this accretion process, Jupiter itself was forming slightly further out from the Sun. Jupiter's gravitational pull disturbed the orbits of the pieces of protoplanet in the asteroid belt, changing them from circular paths more or less parallel to one another to a confusion of individual elliptical orbits, criss-crossing each other, as if the runners were veering at full speed across all the lanes in the track as they ran. With this pattern of orbits, the lumps of rock that had by now formed in the asteroid belt began to collide with one another at much higher speeds – so high, in fact, that instead of sticking together they smashed each other apart.

Because some of the meteorites that reach the Earth, and which presumably originate in the asteroid belt, are highly metallic, it seems that some of the protoplanetary pieces must have got large enough to form molten interiors with metallic cores. Hughes calculates that altogether about eight asteroids as big as Mars should have formed during the accretion phase in the asteroid belt. On this picture, Mars itself might be regarded as a leftover asteroid from that era, and not a planet in the same sense that the Earth and Venus are planets. There should also have been about 640 asteroids bigger than Ceres, before the break-up began.

All of these large objects broke up in the collisions caused by the disturbing influence of Jupiter (except, possibly, that one of the Mars-sized asteroids may have fallen into the inner Solar System and collided with the Earth in the Big Splash that created the Moon). This whole process of accretion followed by collisions and fragmentation may only have taken about a hundred million years or so. Under the disturbing influence of Jupiter, out of the four Earth masses of material originally present in the belt, all but 0.3 per cent of an Earth mass (one twelve-hundredth of the original total) has been lost as pieces of debris have been disturbed into orbits that either take them to a fiery death in the Sun, or out of the Solar System altogether.

The influence of Jupiter on the orbits of objects in the asteroid belt can still be seen today, in the form of the so-called 'Kirkwood gaps'. As their name implies, these are regions in the asteroid belt where very few asteroids are seen – orbits which seem to have been swept clean. They were described and explained by Daniel Kirkwood, in the middle of the 19th century. The gaps occur for orbits which correspond to a simple fraction of the period of Jupiter's orbit – where an asteroid would orbit in 1/2, or 1/3, or 1/4 or 1/5 of the time it takes Jupiter to orbit the Sun once, and so on. In such an orbit, an asteroid would receive a rhythmic tug from Jupiter, which would build up in the same way that a small rhythmic push on a child's swing can set it moving through a very large arc. The missing asteroids from the Kirkwood gaps have been tugged into very different orbits – again, some of them may have escaped from the Solar System, others may have collided with their neighbours (although this is very unlikely today, since there are huge gaps between individual asteroids), and some may have tumbled into the inner part of the Solar System. This is where things get interesting for us.

In 1873, an American astronomer, James Watson, discovered an asteroid, dubbed Aethra, which has an orbit that brings it closer to the Sun than Mars is, before swinging back out into the main asteroid belt. In 1898, the German Gustav Witt found another asteroid, Eros, which has a closest approach to the Sun of 1.33 AU, just 22 million km outside the Earth's orbit. In 1932, the Belgian Eugene Delporte discovered Amor, which grazes past just outside the orbit of the Earth around the Sun, at a closest distance of 14 million km to the Earth's orbit, just 1.08 AU from the Sun. Then, a few weeks later, Karl Reinmuth, working at the Heidelberg Observatory, found an asteroid now known as Apollo, which actually crosses the Earth's orbit and has a closest approach to the Sun 50 million km inside the orbit of the Earth. In May 1932, Apollo passed within 12 million km of our planet.

Since the 1930s, several more asteroids that come near, or cross, the Earth's orbit around the Sun have been identified. The ones which spend most of their time further out from the Sun than us, but which cross the Earth's orbit when they come closest to the Sun, are named the Apollo asteroids, after the first of their kind to be found. The so-called Aten asteroids (named after an object discovered in 1976) have the opposite pattern of behaviour; they spend most of their time closer to the Sun than we are, but cross the Earth's orbit when they are at their furthest from the Sun. And the Amor asteroids, like their archetype, have orbits which come close to but do not actually cross the Earth's orbit (in round terms, they approach to within 1.3 AU of the Sun). There are only three known Aten asteroids, 23 Apollos and 20 Amors. But such objects can only be seen when they come close to us, and the statistics suggest that there must be more than a thousand Earth-crossing asteroids (at least 1,300) bigger than 1 km in diameter – certainly big enough to be noticed by astronomers the next time they do come close.

How big is that? Apollo itself has a diameter of about 1.4 km, typical of the multitude of ordinary asteroids found in the asteroid belt. When Eros passed in front of a star on 23 January 1975, careful observations of the way the tumbling cosmic rock made the light from the star flicker on and off showed that the asteroid is roughly brick-shaped, 30 km long, 19 km wide and 7 km deep. And how close do they come? In 1937, Reinmuth discovered an object now known as Hermes, about a kilometre across, which passed within 700,000 km of the Earth, less than twice the distance to the Moon. And on 14 June 1968 another asteroid, Icarus, passed within 6.4 million km of the Earth.

Obviously, some asteroids come even closer than that; we don't have to look any further to find the source of the cosmic impacts that have occurred over the past few hundred million years. A collision with an asteroid 1 km across would create a crater 13 km in diameter, with the explosive force of 20,000 Megatonnes of TNT; and the statistics suggest that such a collision should happen, on average, every million years or so, exactly matching the cratering record of the past few hundred million years on Earth and the other inner worlds. Although there may be only a thousand or so of these Earth-crossing objects around at any one time, as they are taken away by collisions with the Earth (or some other planet), the supply is replenished by new escapees from the main asteroid belt.

Most of the Earth-crossing asteroids are indeed escapees from the asteroid belt, disturbed into more elongated, elliptical orbits by the tug of Jupiter. Some, though, may be dead comets, masquerading as asteroids. Which brings us to the second source of cosmic impacts on Earth.

People have known about comets for as long as they have looked at the skies. Indeed, our distant ancestors probably knew more about comets, in one sense, than the average person does today. Comets can be spectacular sights in the night sky, with tails that stretch literally half way across the sky, corresponding, in some cases, to a real length greater than the distance from the Earth to the Sun. Hardly surprisingly, these heavenly visitors, appearing unpredictably, were seen by our ancestors as signs from the gods, and often as harbingers of doom. Judging by what happened at Tunguska almost a hundred years ago, they may have had good reason to link comets and disasters.

In modern times, though, comets are less familiar to most people, as the spread of city lighting makes it harder to get a decent view of the night sky. And, coincidentally, in recent decades there have been relatively few bright comets to see, and none of the really spectacular variety. So while the scientific understanding of comets has progressed enormously since about 1950, this exactly coincides with a period when comets have ceased to be a natural phenomenon that most people can see for themselves.

The first step towards an understanding of comets was made by Edmund Halley at the beginning of the 18th century, and published in 1705. He used Isaac Newton's discovery of the relationship between the law of gravity and the orbits of objects moving around the Sun to calculate the orbits of 24 comets. He realised that what people had thought to be four different comets appearing in the years 1456, 1531, 1607 and 1682 were actually the same object moving around the Sun in accordance with Newton's law of gravity.

The difference between the orbit of the comet (now known, of course, as Halley's Comet) and the orbits of the planets is that planets move in nearly circular orbits, while the comet moves in an elongated elliptical orbit, taking it far out from the Sun at one end of the orbit (in the case of Halley's Comet, out beyond Neptune), but close to the Sun at the other end (in the case of Halley's Comet, closer than Venus, at about 0.58 AU). The extensive tail which makes the comet easily visible only grows when it comes in to the inner part of the Solar System and heats up as it gets closer to the Sun, which is why nobody had observed it travelling round its entire orbit. Halley predicted that the comet would return again in 1758, which it did, 16 years after his death.

Halley's comet is still travelling around the same orbit, and returned to our part of the Solar System most recently in 1986. It wasn't a spectacular sight then, because the Earth was in the wrong part of its own orbit to make the comet easily visible. The next return, in 2061, is also likely to be unremarkable.[1] But in March 2134 our descendants should be in for a treat, when Halley's Comet passes within 14 million km of the Earth, less than half the distance at which it put on an impressive show back in 1910.

The key feature of Halley's investigation of comets was that he tamed them, showing that they were not maverick wanderers of the skies or messages from the gods, but subject to the same scientific laws as everything else in the the Solar System – indeed, everything else in the Universe. But while all comets were seen to be obeying Newton's law of gravity and his laws of motion, as astronomers applied the techniques pioneered by Halley to more and more comets it became clear that they were dealing with two different families of these objects.

Halley's Comet itself is the archetypal example of a 'short-period' comet, which travels in an orbit stretching across the region of the Solar System in which the planets orbit, and which returns close to the Sun once every few decades. The orbits of these comets may shift slightly as a result of gravitational perturbations by the planets, but they are easily identifiable as the same objects coming round and round again. They don't all go out as far as Halley's Comet. Encke's Comet, for example (which had its orbit calculated by Johann Encke in the 19th century), moves between 0.34 AU and 4.08 AU from the Sun, never getting as far out as Jupiter, in an orbit with a period of 3.3 years. Most short-period comets, though, do have orbits which go out about as far as the orbit of Jupiter, indicating that they are to some extent under the influence of the largest planet in the Solar System. Two-thirds of all periodic comets are in orbits that extend no further out from the Sun than 1 AU beyond the orbit of Jupiter.

But there are some comets with highly extended orbits, which come into the inner Solar System from far beyond the orbit of Neptune, swinging past the Sun and back out into the depths of space again. There orbits are so long that it will take them millions of years to travel once around the Sun. These are the 'long-period' comets, many of them first-time visitors to our part of the Solar System, coming from far out in space. Entirely arbitrarily, the formal distinction between a short-period comet and a long-period comet is set at an orbital period of 200 years. This is a little misleading. The really interesting distinction is between comets that have periods of even a few hundred years and those that have periods of millions of years, or which come into our part of the Solar System on such long, thin orbits that it is impossible to calculate a precise orbital period at all. So far, since Halley's ground-breaking study astronomers have measured accurate orbits for rather more than a hundred short-period comets and more than 500 long-period comets; the exact numbers go up each year as more comets (about five a year) are discovered.

At the beginning of the 19th century, the French polymath Pierre Simon Laplace speculated that Jupiter's family of short-period comets might have been captured by the giant planet from the stream of 'wild' comets coming in from the depths of space. Half a century later, the idea was taken up by the American mathematician and astronomer Hubert Anson Newton (no relation to Isaac), who calculated that only one very long-period comet in a million would be captured by Jupiter into an orbit with a period comparable to, or less than, Jupiter's own orbital period of 11.86 years. One in a million may be long odds, but if enough comets kept coming in from the depths of space Jupiter's gravitational attraction would certainly account for the existence of the pool of short-period comets.

But there is another side to the coin. It is much harder to capture a comet flying in on a long thin orbit than it is to give it a gravitational nudge in such a way that it escapes from the Solar System altogether, never to return. In the early decades of the 20th century, astronomers realised that nearly half of the very long-period comets coming in past the planets would be disturbed in this way, and be lost forever by the Sun. Even the 50 per cent or so that stay in their long orbits will be slowed slightly on each passage, ending up in orbits with slightly shorter periods, eventually becoming the kinds of comets we see today with orbital periods of hundreds or thousands of years. So there must be a source of wild comets somewhere far out in space, a reservoir from which a steady trickle of comets escapes into our part of the Solar System. Either our Galaxy is swarming with interstellar comets, which are captured from time to time by the Sun's gravity, or there has to be a cloud of comets, far out from the Sun, but still held in place by its gravitational grip. Because comets come from all directions, the cloud would have to be like a spherical shell of material, completely surrounding the Solar System.

At first, few people took the comet cloud idea seriously. But in 1932 the Estonian Ernst Öpik, then working at Harvard, tried to calculate the maximum possible size of the hypothetical comet cloud which the Sun could cling on to. He estimated that even at a distance of 60,000 AU (1,500 times the average distance of Pluto from the Sun) comets would be held by the Sun's gravity, safe from being swept away into space by disturbances from passing stars for the entire history of the Solar System.

At that time the existence of this cloud was still a speculation, an inspired guess. It still seemed possible that some (or all!) of the long-period comets might be visitors from interstellar space, and had never been part of the Sun's family before they plunged in on their extended orbits. The only way to check the idea was by painstakingly calculating the orbits of very long-period comets accurately, and tracing them back to their origins. This was a laborious task in the days before electronic computers. By 1950, there was enough orbital data available for the Dutch astronomer Jan Oort to come up with an improved picture of the way that comets fit in to the Solar System. He studied 21 long-period comets, and found that most of them had orbits which started out around about 100,000 AU from the Sun. Since 1950, many more comet orbits have been studied, with more powerful computers that can make accurate orbital calculations quickly, and these studies have confirmed Oort's picture. Most long-period comets come in to the inner Solar System from a vast cloud of comets surrounding the Sun at a distance of about 100,000 AU, literally half way to the nearest star.

The cloud is commonly known as the Oort Cloud, and by more scrupulous astronomers as the Öpik–Oort Cloud.

A typical comet in the cloud is orbiting around the Sun at a leisurely 100 metres per second or so (rather faster than 320 km per hour), and may have done so for billions of years, since the Solar System formed. Just occasionally, the gravitational disturbance caused by the influence of a nearby star will send a few comets falling gently inwards, speeding up as they fall. Until, millions of years later, one or more of them hurtle through the inner Solar System, whip round the Sun at high speed like a car in the Indianapolis 500 navigating the steep banking flat out, and speed off back out into space. And just one in a million of these visitors from the Öpik–Oort Cloud passes close enough to Jupiter for the giant planet to grab hold of it gravitationally and swing it into a short-period orbit around the Sun. But how did the comets of the Öpik–Oort Cloud get there in the first place?

Most astronomers agree that it really was 'in the first place' – at the time the Solar System was born and the planets formed. But there is a slight disagreement about exactly how the Öpik–Oort Cloud got filled up with comets, its extent, how many comets there might be in the cloud and how big they are. First, here's the standard explanation, as taught to student astronomers today.

When the planets were forming in the primordial disc of material around the young Sun, most of the volatile material, such as ice, in among the dust grains in the inner Solar System was evaporated by the Sun's heat, which is why the four planets nearest the Sun are small and rocky. Out beyond the orbit of Jupiter, however, it was cold enough for ice to stay frozen – not just water ice, but frozen ices containing other material, such as methane and ammonia. So the basic building blocks from which the outer planets formed contained a wealth of material that would have been evaporated had it been closer to the Sun. This is why the outer planets are huge balls of gas. But there must have been far more material in the region beyond Jupiter than the amount of stuff that went in to the other three giant planets. The region between Jupiter and Neptune must have been full of icy dustballs, which came under the gravitational influence of the growing giant planets. These icy dustballs were the original cores of comets, their nuclei.

Some of them were disturbed into orbits taking them close by the Sun, where they evaporated after a few orbits. But others were ejected outward in a kind of gravitational slingshot effect, taking them (eventually) about 100,000 AU from the Sun. Out there, moving very slowly at the furthest ends of their orbits, they came partially under the influence of other stars, tilting and altering their orbits in random fashion, and gradually disconnecting them from the influence of the planets until they became a spherical swarm around the Sun – the Öpik–Oort Cloud.

Detailed analysis of cometary orbits now shows that there are several components to the cloud. As well as the outer cloud, which has an estimated population of about a trillion comets (that is, a thousand billion, ten times the number of stars in the Milky Way), there is an inner cloud, extending out only a few thousand AU from the Sun, which contains six times as many comets – the numbers are required to match the observed rate at which 'new' comets appear in the inner Solar System. The inner part of this inner cloud is not spherical, but flattened in towards the plane in which the planets orbit the Sun. Finally, just beyond the orbit of Neptune lies a region known as the Kuiper Belt, a fat disc containing at least a billion comets. The existence of this trans-Neptunian belt of comets was first proposed by Gerard Kuiper (who was born in Holland but emigrated to the US in 1933) in 1951, at about the time that the idea of the Öpik–Oort Cloud was becoming popular. He suggested that comets might have formed not just in the region where the giant planets formed, but in a primordial disc extending as far out as 100 AU from the Sun. As we shall see, this may well have been too modest a proposal, but it was a new and exciting idea at the time.

In all, there are at least ten trillion and possibly a hundred trillion comets in the outer part of the Solar System, beyond the orbit of Neptune. In cross-section, the whole structure would look like the flaring mouth of a trumpet, opening up from the disc of the Kuiper Belt through the inner cloud and out to the Öpik–Oort Cloud proper. But the whole cloud is so big that it will take the Voyager spacecraft, now moving out of the Solar System after the close encounters with the giant planets, ten thousand years to reach its outer limits.

One of the most important features of this standard model is that it sees comets as a natural product of the formation of the Solar System (and of planetary systems in general), along with planets and asteroids. David Hughes has calculated that out of an original 120 Earth masses of material in the Saturn–Uranus–Neptune region of the young Solar System, 85 Earth masses went to form the cores of these three giant planets, and 35 Earth masses was left over in the form of cometary nuclei. Half of these comets were perturbed into orbits taking them close by the Sun (some of these may have hit the inner planets, perhaps contributing their atmospheres and water), and of the rest no less than 16 Earth masses of material has been completely ejected from the Solar System during its long history.[2] Which means that the entire mass of all the billions of comets left in the Öpik–Oort Cloud today is less than 1.5 times the mass of the Earth. Allowing for uncertainties in the calculation, Hughes is willing to double this number and say that the cloud contains no more than three times as much mass as the Earth.

On this standard picture, the nuclei of comets are fluffy snowballs, with densities about a fifth of the density of water; and most astronomers accept the guesstimate that the largest members of the Öpik–Oort Cloud are no larger than a tenth of the size of the core of a planet like Saturn – about as big as the Earth. This image of the core (nucleus) of a comet as a 'dirty snowball' was developed almost exactly in parallel with the development of the idea of the comet cloud itself, following a suggestion published by the American astronomer Fred L. Whipple (no relation to the British meteorologist F. J. W. Whipple) in 1949. The key difference between asteroids and comets is that comets have snow, while asteroids do not. But the rest of what comets have is much the same as what went into rocky asteroids, although, of course, no cometary material has ever been part of the molten core of a Mars-sized asteroid.

In his book *The Mystery of Comets*, Whipple describes how he came up with the dirty snowball idea. In the 1940s, spectroscopic studies showed that the heads of comets (the extended balls of gas which grow around the nucleus as the comet approaches the Sun) contained many molecules which seemed to be derived from water, methane, ammonia and carbon dioxide. There had to be a large reservoir of these molecules in order to keep comets active for hundreds, or even thousands, of revolutions around the Sun; and it seemed obvious to Whipple that 'the nucleus of a comet must have a great mass of ices embedded with dust or meteoric particles – in other words, it must be a huge, dirty snowball.'[3] In a vacuum, ices do not melt when they are heated from outside by energy from the Sun, but simply turn into gas at the warmed surface – they sublimate, passing straight from solid to vapour. There is nothing particularly mysterious about this. Snow often sublimates on a dry, sunny day here on Earth, disappearing almost literally before your eyes, seemingly without trace; and frozen carbon dioxide ('dry ice') also sublimates at room temperature and pressure.

As Whipple acknowledges, the idea that comets are made of ice actually goes back to Laplace who, as we saw, also originated the idea of a cloud of comets around the Sun. But nobody took the idea seriously until the 1950s because astronomers discovered that comets, in many cases, move in the same orbits as streams of particles that produce meteor showers when they enter the Earth's atmosphere. These meteor streams were obviously made of dust; so it was assumed, until Whipple came along, that comets themselves were largely made of dust and rocky particles rather like gravel.

In the 1950s, the spectroscopic evidence was compelling, and the dirty snowball model quickly became accepted. It matched up nicely with the growing realisation that comet nuclei originally formed in the cold depths of space, possibly even beyond the orbits of the giant planets. And simple calculations showed that there really could be enough icy material in a comet nucleus to keep it active for a long time.

Take the example of comet Arend–Roland, which passed about 0.32 AU from the Sun in 1957. At its closest approach to the Sun (perihelion), the comet was giving off about 10 tonnes of gas every second, carrying with it even more dust. This adds up to a total of 2 million tonnes of material lost each day, and overall the comet lost about 100 times this much material during its 1957 visit to the Sun. It seems a vast amount of matter by any human standards. But if the original dirty snowball of this particular comet was just 6 km across, on this visit to the Sun it would have lost only the top metre or so of its surface. Clearly, the reservoir is big enough to keep the comet active for many orbits around the Sun; and, to the delight of Whipple and his colleagues, the amount of ice sublimated in this way on this particular visit by the comet closely corresponds to the maximum amount that could be turned into vapour by the amount of heat the comet is calculated to have received while it was near the Sun.

Eventually, of course, all the ice will disappear, leaving the dust and rocky debris from the original cometary nucleus. The dust gets spread right around the comet's orbit, as a meteor stream, and if there are large lumps of rock left behind then they will look for all the world like asteroids. Some, and possibly many, of the Apollo asteroids could have originated in this way. Short-period comets are a likely source of new Earth-orbit crossing asteroids.

Comets can even develop a rocky core surrounded by a layer of ice, exactly like a frozen asteroid, if the ice melted long ago when the Solar System was young. This is far from being a wild idea, because it is thought that radioactive material spewed out by the supernova explosion of a nearby star was present in the collapsing interstellar cloud from which the Solar System formed. Indeed, many astronomers think that it is the explosion of a supernova, the death of an old star, which triggers interstellar clouds to collapse and form new stars.[4] The radioactive heat produced by this material would have been able to melt proto-cometary nuclei a few tens of kilometres (or more) across, allowing them to segregate into rocky core and icy outer layer as the radioactivity dissipated and the ices froze again. Ice warmed in this way by radioactive energy inside the comet does indeed melt, rather than sublimating.

After all of its icy outer material has been worn away by the heat of repeated passages past the Sun, such a layered object ends up looking exactly like a wandering lump of rock that had been ejected from the asteroid belt. An alternative process, which (as we shall soon explain) seems to be at work in the nucleus of Halley's Comet, is the possibility that the icy core of a comet might end up completely covered in a layer of dark, rocky material. This dust 'mantle' forms from particles that are too heavy to be carried away by the gas from sublimating ice, and could become thick enough to provide an insulating blanket around the ice, stopping any more sublimation taking place. The comet would have effectively disguised itself as an asteroid. Intriguingly, observations with powerful telescopes show that several of the asteroids that come near to the Earth's orbit have very dark, reddish surfaces, a colour typically associated with the kind of organic material (CHON compounds) known to be present in comets.

Either way, there are many good examples of asteroids that may be former comets, including an asteroid known as Phaethon, which follows the same orbit as the meteor stream that produces the meteor shower known as the Geminids, visible each December when the Earth passes through the stream. Meteor showers are, as we have mentioned, associated with comets; but there is no live comet visible in the same orbit as the Geminids. The

finger of suspicion points firmly in the direction of Phaethon as a dead comet that produced the dust which causes
the Geminids before it died. And if comets can turn into,
or disguise themselves as, asteroids, that helps to explain
the source of new Earth-orbit crossing asteroids. Astronomers are having to re-think their definitions and classi-
fications, with the distinction between asteroids and
comets getting decidedly blurred. But the details of this
classification need not worry us too much. The exact
origins of a lump of cosmic debris do not matter if one
of them strikes the Earth – just as much damage will be
caused by a kilometre-sized chunk of dead comet as by a
kilometre-sized chunk of former asteroid.

The basic picture of the nucleus of a comet as a dirty
snowball was confirmed in March 1986, when several
spacecraft flew close by the nucleus of Halley's Comet
during its latest pass by the Sun. One of them, Giotto, sent
back pictures until it was partially disabled by a collision
with a fragment of cometary debris just 605 km from the
nucleus. The images sent back, and other data, showed
that the nucleus is indeed a lump of icy material, with 80
per cent of its mass in the form of water ice. But the
observations also threw up a few surprises, suggesting that
the best image of a comet nucleus might not be so much a
dirty snowball, but an icy rock pile – frozen water (and
other stuff) with lumps of solid material embedded in it.

The nucleus of Halley's Comet is an irregular potato
shape, roughly 16 km long and 8 km across (about the size
of the island of Manhattan). The surface is extremely dark
(usually described as being blacker than coal, which is
literally true) and reflects less than 4 per cent of the light
that falls on it. Because it reflects so little light, the nucleus
is hard to see, and since astronomers had assumed that it

112

had a higher reflectivity, before the spacecraft observations they had not realised how large the nucleus is. The black surface is a crust of dark material (dust laced with organic compounds) on the surface of the ice, which is heated during the close approach to the Sun. This heat penetrates to the icy interior of the nucleus, where the ice sublimates, producing gases that force their way out through cracks in the crust in the form of jets and plumes that carry fine grains of dust away into the tail of the comet. But when the nucleus is far away from the Sun, none of this activity takes place, and all astronomers can see, even with their best telescopes, is the rocky crust of the nucleus, looking like an Earth-orbit crossing asteroid.

The dusty material associated with the comet itself contains a wealth of the life-giving CHON compounds, reinforcing the idea that ancient comets may have played a part in helping life to get started on Earth.

About the same time that Halley's Comet was being studied in such detail, the reflectivity (or albedo) of several other short-period cometary nuclei was determined using an infrared technique. These albedos all turned out to be much lower than anticipated, but in line with the observations of the nucleus of Halley's Comet. If this low albedo is typical of all comets, as now seems likely, all the estimates of the sizes of their nuclei based on brightness measurements have to be increased. This, and other observations of Halley's Comet itself, have important implications for estimates of the total mass of material in the Öpik–Oort Cloud, and of the maximum masses of individual comets. The details were spelled out by Leonid Marochnik, Lev Mukhin and Roald Sagdeev, of the Space Research Institute of the Academy of Sciences in Moscow, shortly after the Halley missions.[5]

The way in which the trajectories of the spaceprobes were disturbed as they flew past Halley's Comet shows that the nucleus has a mass of 300 billion tonnes. The Moscow team argue – on the evidence of the new albedo measurements and analysis of the orbits of Halley's Comet and other short-period comets – that this can be taken as a lower limit for the mass of what they call an 'average new' comet coming from the Öpik–Oort Cloud. They go on to make the case that the size of Halley's nucleus is also typical of new comets coming in from the cloud. This means that estimates of the total mass in the cloud have to be increased dramatically.

David Hughes' estimate of the total cometary mass which ended up in the Öpik–Oort Cloud after the birth of the Solar System is, if anything, a little on the low side compared with the value accepted by most astronomers before the Halley missions, which was several times the mass of the Earth. But in the light of the Halley data, there must be a minimum of 100 Earth masses of material in the form of comets in the cloud proper – a quarter as much mass as there is in all the planets of the Solar System put together. And there must be even more material in the flaring disk of the extended Kuiper Belt.

This amount of material is too large for all of it to have got into the cloud by being ejected by gravitational interactions from the region in which the giant planets formed. Some of the comets may indeed have been ejected in this way; but there is a growing view among astronomers that the primordial disc of material around the Sun extended far beyond the orbit of Neptune, and that many millions of comets formed originally very far out from the Sun.

Studies of the extended disc of material around the young star Beta Pictoris lend weight to this idea. That disc extends out to about 1,000 AU from the star (more than 30 times the distance of Neptune from the Sun) which is thought to be about 200 million years old, and has about one and a half times the mass of the Sun. Discs have also been detected around other stars, although only the Beta Pictoris disc has yet been photographed.

It seems that astronomers may have been too parochial, too 'planet chauvinist', in working out the earlier models of how comets formed, and that they should not have ignored the outer regions of the primordial disc. Mark Bailey, of Liverpool John Moores University, has found that if comet formation is modelled in a computer, on the assumption that there was a disc more like that seen around Beta Pictoris present around the young Sun, then about 400 Earth masses of material, roughly the same as the amount of mass in all the planets, ends up in the form of comets. The 'extra' comets are primarily located in a more massive, more extended inner disc, a kind of super Kuiper Belt, linking the planetary part of the Solar System to the outer Öpik–Oort Cloud. It could be argued, in fact, that this is now a conservative estimate, because the Moscow team would place the figure for the total mass in the disc linking the planets to the Öpik–Oort Cloud at higher than 1,000 Earth masses!

This idea is backed up by new spectroscopic studies of the composition of comets. The presence of chemical compounds such as carbon monoxide in large quantities in the gas escaping from comets when they are far from the Sun shows that they formed under extremely cold conditions – where the temperature was less than –250 °C. This immediately pushes the zone of formation of these comets out to beyond 100 AU from the Sun, three times further out than Neptune.

There is another implication of these studies. Computer simulations, like those carried out by Mark Bailey, tell us that the largest individual objects formed in the extended cometary disc could be as much as 300 km across – supercomets, like nothing seen in the inner part of the Solar System since records began. Such objects would certainly not be delicate fluffy snowballs, dirty or not. They would have been big enough to melt through radioactive heating when the Solar System was young, and to segregate out rocky cores – which could themselves contain boulders tens of kilometres across – surrounded by layers of solid ice. If an object like that fell into the inner part of the Solar System, it could wreak havoc on a scale greater than anything we have discussed yet, except for the Big Splash in which the Moon was formed. The bad news is that such supercomets are not a figment of the astronomers' imagination, nor simply an artefact of the computer simulations. They have actually been seen, although at first nobody realised what they were. And it now seems likely that the break-up of such objects in the inner Solar System could represent the biggest long-term threat to the environment here on Earth.

Notes

1 The interval between returns can change slightly, as the orbit of the comet is altered by the gravitational pull of the giant planets when it is in their part of the Solar System.

2 On these figures, 12 comets were lost into space for every 1 that went into the Öpik–Oort Cloud. If other stars ejected comets at the same rate, there may be ten million billion billion comets in interstellar space, and once in every century or so one of them might plunge into the Solar System. But none of the long period

comets observed so far can be unambiguously identified as one of these interstellar interlopers.

3 *Mystery*, p. 147.
4 See *In the Beginning*, by John Gribbin.
5 *Science*, volume 242, p. 547, 1988.

VI

BREAKING UP IS NOT SO HARD TO DO

Since 1989, astronomers have been forced to make dramatic upward revisions to their estimates not just of the number of comets in the outer part of the Solar System, but also of the size of the largest comets. Until then, it was generally accepted that the nucleus of a comet – the dirty snowball itself – would never be much bigger than those of the short-period comets we see in the inner Solar System today. That meant a typical comet nucleus would be only a few kilometres across, and even out in the Kuiper Belt or the Öpik-Oort Cloud the largest comets might be no more than a few tens of kilometres in diameter. But in 1989 an object some 200 km in diameter (some observations suggest it may be 300 km across), previously identified as an asteroid, was seen emitting gas and growing a fuzzy coma, a cloud around its nucleus. Since then, the same object has been seen emitting jets of gas as it has got closer to the Sun. It is, without doubt, a comet. But in this case, fortunately, 'closer' to the Sun still doesn't bring the object, called Chiron, in to our part of the Solar System.

Chiron was discovered by the American astronomer Charles Kowal in 1977. It orbits from 8.5 AU to 19 AU from the Sun, between Jupiter and Uranus, crossing the orbit of Saturn. When it was discovered, some eager reporters optimistically described it as a 'new planet', but astronomers classified it as an asteroid, and speculated that it might be one of the larger members of an outer asteroid belt. Luckily though, at the time it was discovered, Chiron, which has a 51-year orbit, was moving towards its closest approach to the Sun, which it reached, just inside the orbit of Saturn, in February 1996.[1] The modest heating it received as a result was enough to reveal its true nature: a giant comet.

The first sign of something unusual came in 1988 when, at a distance of about 12 AU from the Sun, Chiron doubled in brightness. Since then, it has been continuously active in one way or another, at such large distances from the Sun that the activity must be driven not by the sublimation of water ice but by ices made of carbon monoxide, methane, and possibly solid nitrogen. There is certainly plenty of material to provide the activity: Chiron is about 40 times bigger in diameter than what used to be thought of as a typical comet, and about 65,000 times as massive. With one giant comet identified in the outer Solar System, in the 1990s astronomers eagerly, and successfully, searched for more.

One important feature of the orbit of an object like Chiron, passing near to the orbits of the giant planets (and actually crossing the orbit of one of them) is that it is unstable, in the long term. In fact, apart from one special case, all orbits are unstable, but some are more unstable than others.

The special case is the orbit of a single planet around an isolated star. Under those conditions, the law of gravity and the laws of motion described by Isaac Newton apply perfectly, and the planet follows the same precise orbit forever. The equations that describe the orbit are said to be 'integrable', and they can be used to calculate the position of the planet in its orbit at any time in the past or in the future. But as soon as you add a third object to the system, things change. The 'three body problem', as it is known, is not integrable (nor, of course, is the situation for more than three objects). This means that it is impossible, even in principle, to calculate the orbits of each of the three bodies into the indefinite past or the indefinite future.

Of course, in a system like the Solar System you can come pretty close to predicting the orbits of the planets. The Earth is not, for example, suddenly going to swing off in some new direction through space. The orbits of the planets are stable on any human timescale, because the Sun is so big compared with the rest of the objects in the Solar System. The way you can calculate the orbits when there are two (or more) planets orbiting a star like the Sun is straightforward but tedious. Because the Sun is so big, you can ignore the way it moves as it is tugged by the relatively feeble gravitational pull of the planets. Then, you imagine holding all of the planets except one still, and calculate a small part of the orbit for that planet, starting from its present position, under the combined gravitational pull of all the other planets and the Sun. But by moving one planet a little bit, you change that planet's gravitational influence on all the others. Now you have to hold *that* planet still in its new position, and allow one of the other planets to move on a tiny bit in its orbit; and so on through a tedious sequence of repeated calculations.

The smaller you make the move allowed by an individual planet at each step of the calculation, the more accurate a prediction of the orbits you will get – and the longer and more tedious you will make the calculation. But because in reality all the planets (and the Sun) are moving all the time, it is impossible to make a perfect prediction of how any object in the Solar System will move. Indeed, there is no such precise prediction – the planets, asteroids and comets do not themselves 'know' exactly where they will be at some time in the distant future.

Until the advent of high-speed electronic computers able to run these kinds of calculations for huge numbers of steps, nobody was sure just how stable orbits in the Solar System are. But now the calculations have been carried out, and we know the answers. By and large, it turns out that the orbits of the planets are indeed stable for billions of years – vastly longer than any timescale we have to worry about. But for smaller objects, and especially for smaller objects that pass close by large planets, the orbits become unpredictable on a much shorter timescale, and are subject to what the mathematicians call chaos.[2] This means that at some point in its evolution, the orbit of an object like Chiron changes suddenly from its present 51-year orbit to something else. The something else will probably shift Chiron itself into a new orbit taking it out of the Solar System; but other objects in quite similar orbits would switch the other way, plunging closer in to the Sun, where they would stay for a while before (if they avoid collision with one of the inner planets) switching into yet another orbit, and so on.

The point about chaos is that the precise 'new' orbit is unpredictable, even though the fact that the object will change orbits is certain. It happens because the details of the process depend on the *exact* influence of all the planets on (in this case) Chiron, and that influence cannot be calculated precisely. As an example of the importance of a tiny difference in some starting conditions making a huge difference in the end conditions of a process, think of a marble balanced on top of a perfectly smooth, rounded dome. We know that the marble will roll off in one direction or another, but there is no way to predict which direction it will roll in. And if it is not quite dead-centre on top of the dome, a tiny difference in exactly where the marble is placed on the top of the dome can make a big difference in where it ends up after it has rolled down. In the same way, a tiny difference in the influence of the planets on Chiron will make a big difference in the orbit it ends up in.[3] So although computer simulations show that Chiron is in an unstable orbit and that it cannot stay there for more than about a million years, no computer simulation will ever be able to say in advance exactly when it will fall out of this orbit, or exactly which new orbit it will fall into.

But that doesn't really matter. What does matter is that those computer simulations show that Chiron cannot have been in its present orbit for as long as a million years. It must have been perturbed out of a previous orbit, further out from the Sun, in the Kuiper Belt itself. So the search for other Chiron-like objects in the early 1990s did not focus on the region between Jupiter and Uranus where Chiron now is, but further out, beyond Pluto, where astronomers believed it must have remained in 'cold storage' for billions of years, since the formation of the Solar System.

Even so, Chiron is not entirely alone. In 1991, astronomers at the Kitt Peak National Observatory, in Arizona, found a similar object, about the same size as Chiron, orbiting between Saturn and Neptune; it is now called

Pholus. A third of these objects, called 1993 HA2, was found a year after the first Kuiper Belt comet proper was discovered by David Jewitt and Jane Luu, working at the Mauna Kea Observatory in Hawaii, in August 1992. They used a sensitive electronic detector called a charge coupled device (CCD), attached to a large telescope, to search for faint objects beyond the orbit of Neptune (which is at 30 AU from the Sun). Because CCDs can only look at a small patch of the sky at a time, the search involved years of careful effort, which resulted in the discovery of an object so faint that the faintest star visible to the unaided human eye is six million times brighter – so faint, indeed, that even CCDs could only pick it out near the time of new Moon, when the background light from the sky was at its faintest.

The object was given the identification label 1992 QB1. The 'Q' is based on a standard classification which corresponds to the second half of August, the month of discovery, and the 'B1' refers to the order in which the object was discovered within that period. So many odd bits of cosmic debris had been identified by astronomers around the world in August that all the letters A to Z had been used up, and they were working their way through the alphabet again. So 1992 QB1 was the 28th object discovered in the second half of August that year (if it had been the 26th, it would have been dubbed 1992 QZ; if it had been the 30th, 1992 QD1, and so on). But astronomers soon dropped the '1992' and took to referring to the object simply as QB1.

Assuming that QB1 reflects about the same amount of light for each square metre of its surface as an ordinary comet, it must be about 200 km across – similar in size to Chiron, and about one-tenth of the diameter of Pluto. Its present orbit keeps it about 44 AU from the Sun, considerably closer than the region where the main Kuiper Belt is thought to be concentrated, beyond about 100 AU out from the Sun. Most probably, it has come in from the Kuiper Belt, having been disturbed in some way, and is in an intermediate orbit, unstable in the long term, from which it may work its way inward, perhaps eventually all the way to the inner part of the Solar System.

Once one Kuiper Belt comet had been discovered, more followed quickly. Five more in 1993, 17 altogether by the end of 1994, and more than 20 by the middle of 1995 (about half of them orbiting more than 40 AU from the Sun). This led to an estimate that there may be at least 35,000 objects larger than 100 km in diameter orbiting in this region of the Solar System, just beyond the orbit of Neptune. To put that figure in perspective, there are only 200 objects that big in the asteroid belt between Mars and Jupiter. Most of these objects are a few hundred kilometres across. But the searches which identified them should have shown up anything larger than about 600 km across by now, and they haven't; which suggests that there may be very few objects that big (if any) in the inner Kuiper Belt. The total mass of these supercomets would be at least 3 per cent of the mass of the Earth, and they should be accompanied by 1–10 billion objects the size of Halley's Comet, with all this representing only the innermost part of the Kuiper Belt.

Some of these objects orbit between about 32 and 36 AU from the Sun, with a gap before the rest are found between 42 and 46 AU from the Sun. Computer simulations show that this entire region, from about 30 AU to 45 AU, is one in which chaos operates and stable orbits do not exist. This tells us both that these supercomets have been fed in to the inner Kuiper Belt from further out, and also that they are destined to move on; some being ejected from the Solar System altogether while others work their way further inward still, eventually ending up in our part of the Solar System.

That, of course, also applies to their smaller companions, and it now seems that this inner Kuiper Belt region is the source of all short-period comets that come in to the inner Solar System, including Halley's Comet itself. All the comets with orbits around 35 AU are quickly 'eroded' in this way, but either side of this region of extreme instability there are two bands of more stable (or, at least, less unstable) orbits where comets can survive for millions of years.

Some astronomers like to refer to these objects as 'ice dwarfs', rather than comets, emphasising their role as the (probable) building blocks from which the giant planets were constructed. They are also intrigued by Pluto, which, although ten times bigger than QB1 (ten times the diameter, that is, which means a thousand times the volume) and in an unusually stable orbit, crosses the inner band of 'stability' beyond Neptune, and might be thought of as simply an extremely large variation on the ice dwarf theme. At least, it might be thought of in that way if it were simply made of ice.

Observations of the orbit of Pluto and its moon Charon (not to be confused with Chiron!) show, however, that Pluto is a rocky planet with an icy covering (75 per cent rock to 25 per cent ice), while Charon is a ball of icy material, 1,190 km in diameter. One popular speculation among astronomers is that the two objects started life independently of one another, and that Charon collided with Pluto in an event similar to the Big Splash which created our Moon. In the impact, Pluto lost a lot of ice but retained its rocky core, while Charon went into orbit around Pluto and gathered up the icy debris. But whatever the exact origins of the Pluto–Charon system, the important, and relevant, point is that objects in that part of the Solar System are not all entirely made of ice. They can have rocky cores as well, even if not such extensive rocky cores as in the case of Pluto's.

The finger of suspicion also points at some of the other moons in the outer Solar System. For example, both Phoebe, a small moon of Saturn, and Triton, a large (2,700 km diameter, slightly bigger than Pluto) moon of Neptune, orbit the 'wrong way' around their respective planets, compared with the other moons. This is a sure sign that they have been captured relatively recently – indeed, Triton is slowly spiralling inward and will one day crash into Neptune. This means that there have been objects as large as Pluto or Triton roaming free in the outer part of the Solar System in relatively recent times. Anything roaming near the giant planets could easily get perturbed into a short-period cometary orbit taking it past the inner planets. But even without invoking comets the size of Triton, an icy comet nucleus a few hundred kilometres in diameter, with a total mass of perhaps a hundred million billion tonnes, has room to contain several rocky nuggets – each capable of doing severe damage to a planet like the Earth if they ever came into contact with it. According to the latest interpretation of the evidence, that may well be the fate of some of those lumps of cometary material.

Mark Bailey, of Liverpool John Moores University, estimates that an object the size of QB1 is likely to wander into our part of the Solar System roughly every million years or so. Calculations of what happens to it then are reshaping geological ideas about the catastrophic history of our planet.

Even before the spectacular interaction between Jupiter and Comet Shoemaker–Levy in 1994, astronomers had begun to realise that for comets, breaking up is not so hard to do. Working with Jun Chen, David Jewitt, the co-discoverer of QB1, studied CCD images of 49 comets, obtained between 1986 and 1993, and found that three of them had broken into two pieces. As the pieces drift apart, they would be too widely separated after just six years to appear on a single CCD image. So the Hawaii team concluded that a typical comet suffers fragmentation at least once every hundred years. Typically, a fragment with about a thousandth of the mass of the whole nucleus is split off in this way, so that the main comet survives for a long time, with the fragments adding to the pool of small 'asteroids' that pass relatively near to the Earth.

It is not entirely clear why comets should break up in this way. One suggestion is that as heat from the Sun works its way inwards it may encourage gas to be released in pockets inside the nucleus, where the pressure builds up until it cracks the ice and breaks a piece off. This process is, obviously, more important for comets that come closer to the Sun, but some comets seem to split long before they get close to the Sun. If a large comet came too close to the Sun, however, it would be fragmented in another way – the same way that Shoemaker–Levy was fragmented in a close encounter with Jupiter – by gravity.

Shoemaker–Levy had already been broken into pieces when it was discovered by a long-established team of comet watchers, Carolyn Shoemaker, Eugene Shoemaker and David Levy, working at Mount Palomar Observatory on the night of 24 March, 1993. The Shoemaker-Levy team were so successful at comet hunting that they had found eight comets previously, and since comets are traditionally named after their discoverers (at least, in the case of 'standard' comets with tails – the tradition seems to have been abandoned for supercomets from the Kuiper Belt) this 'new' comet should, strictly speaking, be referred to as Shoemaker–Levy 9. But everyone knows it as 'the' Shoemaker–Levy comet, and with good reason.

At first sight, the object that they found that March night looked on their photographs of the sky like a smeared out, or squashed, comet. The bizarre nature of the comet was confirmed by photographs taken at Kitt Peak by Jim Scotti. What they had found now looked like a string of beads strung out along a single orbit. Further observations of the string of beads in the weeks and months ahead showed that the orbit they were following was around Jupiter, not around the Sun itself, and that the comet had been swung into that orbit by a close encounter with Jupiter on 8 July 1992. The encounter had been so close, indeed, that the tidal forces tugging on the original comet as it swung past the giant planet had broken its nucleus into the pieces that were now strung out along the new orbit. Jupiter itself has a diameter of just over 140,000 km, and the comet had passed within 120,000 km (less than a Jupiter diameter) of the surface. In terms of the size of the planet, that is equivalent to an object passing within 12,000 km of the surface of the Earth.

By the end of April 1993, there was enough information on the orbit to suggest that the fragments were now heading for a collision with Jupiter in the summer of 1994. Cautiously, the astronomers waited to double check their calculations and make more observations of the orbit of Shoemaker–Levy over the next few months, making absolutely certain of their forecast before announcing it to the world. In the week beginning 16 July 1994, the series of impacts of the score or so fragments with Jupiter duly happened, as forecast and right on schedule, producing one of the biggest scientific media events of the decade; and clogging the Internet as millions of images of the impacts were downloaded from scientific databases to be studied by eager enthusiasts. But long before those collisions actually happened, studies of the fragments of Shoemaker–Levy had provided new insights into the structure of comets, and the way they break up.

Astronomers had already suspected that comets formed in the outer part of the Solar System in a process involving at least two steps. First, tiny grains of material in the interstellar cloud from which the Solar System formed would literally stick together and make larger lumps. Originally, these comet seeds would be scattered throughout the spherical cloud of gas and dust; but when the lumps became big enough they would settle down into the plane of the Solar System, with individual icy lumps being pulled together by gravity to make cosmic 'rubble piles'. On this picture a typical comet nucleus would consist of individual icy lumps, all roughly the same size, held together only fairly loosely. This idea matches perfectly with the way Shoemaker–Levy was split apart into its component pieces, each a few hundred metres to a few kilometres across. The image of a primordial supercomet nucleus is of a flying pile of (rather large) snowballs, ready to be spread out by gravitational interactions with the giant planets. Such a rubble pile would be well able to produce an effect like a stream of machine gun bullets when the fragments piled into one of the moons of the outer planets, producing lines of craters like some that are seen on Ganymede and Callisto – and, indeed, on our own Moon.

The individual fragments of the comet – the primordial snowballs – seem to have been fairly solid objects, judging by the effect they had on Jupiter. Some of the impacts released as much energy as the explosion of 10 million megatonnes of TNT, and one the equivalent of about 250 million megatonnes, 15 billion times the power of the Hiroshima bomb. Plumes from the impact sites reached up thousands of kilometres above the cloud tops of Jupiter. The largest fragment must have been about 3 km across to produce such a dramatic result, suggesting that the original nucleus of the comet, before it fragmented, was about 19 km across (roughly the size of the nucleus of Halley's Comet). Dark scars, like black eyes, were left by some of the impacts, and waves from the impacts rippled through the atmosphere of Jupiter.

Intriguingly, the line of temporary dark spots photographed as the comet fragments struck closely resemble descriptions of spots made several times before the days of photography. Respectable observers who described such features on Jupiter include William Herschel and Giovanni Cassini; and in 1927 the astronomer T. E. R. Phillips drew a row of spots he had fleetingly observed on the surface of Jupiter, too faint to be recorded in a short exposure by the photographic equipment available at the time. All firm evidence, if any were needed, that the events of 1994 were not unique, and that Jupiter (and, by implication, the other planets of the Solar System) had been battered in this way before. But if you want to look on the bright side, spectroscopic analysis of the impact sites after the 1994 events shows the trace of many complex organic molecules, the basic building blocks of life, which were either brought to Jupiter by the comet or created in chemical reactions triggered in the fireball explosions.

Just how lucky astronomers were to be able to observe all of this in such detail was spelled out by Eugene Shoemaker at a meeting held in Baltimore to discuss the observations in May 1995. He calculated that on average a comet as big as Shomaker–Levy 9 is broken up by Jupiter's gravitational field, with the fragments colliding with the planet, about once every 2,000 years. For the 1994 impacts to have occurred just after the repair of the Hubble Space Telescope, when the Galileo space-probe was in position to observe them, during the era that astronomers on Earth had developed efficient infrared detectors, and, as he put it, while the US government was still funding basic research, was, in Shoemaker's words, 'a miracle indeed'.

But Jupiter is far away, and none of the fragments from Shoemaker–Levy 9 was ever any threat to life on Earth (in spite of some hysterical predictions by doomsayers at the time). People on Earth should have been much more concerned about the implications of another fragmented comet that was discovered by the American Don Mach-holz less than a month after the last fragments of Shoe-maker–Levy had crashed into Jupiter. This comet, known as Machholz 2, comes in five pieces, the largest about a kilometre across, all of them following the same orbit, which crosses the orbit of the Earth. There is nothing to worry about in the short term; the pieces of Machholz 2 may cross the orbit of the Earth, but none of them will come closer than 18 million km from our planet over the next hundred years or so. But the discovery does serve to focus our attention back closer to home, as we consider what may happen to giant comets when they wander in to our part of the Solar System, in the light of what we know happened to Shoemaker–Levy 9.

As we mentioned in chapter two, there are several impact features on Earth that seem to have been formed at the same time as the Chicxulub crater, 65 million years ago when the dinosaurs died. They include the Manson crater in Iowa and the two craters in Russia at Kamensk and Kara. If these impacts did all occur at about the same time, it cannot have been a coincidence. They must have been produced by fragments of a single object that had broken up and ploughed into the Earth in the same way that the fragments from Shoemaker–Levy ploughed into Jupiter. But if so, the original object must have been huge – the Chicxulub crater itself is one of the biggest known on Earth, and it would simply have been the largest of several more or less simultaneous impacts. And the Earth is a much smaller target than Jupiter, so some (possibly most!) of the fragments may have missed the planet altogether. The only candidate for the parent body is an object like Chiron or QB1.

Mark Bailey's calculations show that sometimes an object like this will get disturbed by the gravitational influence of Jupiter or Saturn and fall in to the inner part of the Solar System. The orbits that these super-comets end up in are not stable, but subject to chaotic dynamics. Nevertheless, although the computer simulations cannot predict precisely what orbit such an object will end up in, they can tell us that eventually any such object switches into a Sun-grazing orbit, going so close to the Sun that it must get broken up into its constituent pieces – in the same way that Shoemaker–Levy was broken up by the tidal influence of Jupiter (quite apart, of course, from the possibility of fragmentation caused by outgassing or some other process). The result is a stream of debris orbiting around the Sun and crossing the orbit of the Earth.

All of this takes place in an interval of less than a million years. So the Earth should have come under the influence of such streams of cosmic debris many times during its long history. Could this explain the repeated mass extinctions of life shown up by the geological record? Some astronomers think so. And there is even evidence that we are living in the aftermath of the break-up of a giant comet in the inner Solar System – a scenario spelled out in impressive detail by Victor Clube, of the University of Oxford, and Bill Napier, of the Royal Observatory, Edinburgh.

Clube and Napier make their case by arguing backwards, studying the orbits of various items of cosmic debris seen in Earth-crossing orbits today, and working out where they came from. The main object that they are interested in is a short-period comet which, like Halley's Comet, is not named after the person who discovered it, but after the first person to work out its orbit and predict its return. What is now known as Encke's Comet, after the German astronomer Johann Franz Encke, was first recorded in 1786 when the French astronomers Pierre Mechain and Charles Messier noticed it as a faint object just visible to the unaided human eye for two nights. In 1795, Caroline Herschel, in England, noticed what we now know to be the same object, but saw it so briefly that, like Mechain and Messier, she did not have enough information to calculate its orbit. It was seen again, equally briefly, in 1805. But the breakthrough came in

1818, when the greatest comet hunter of all time, Jean Louis Pons, observed the comet over a period of seven weeks.

Pons worked as a labourer until he was 28 and then, in 1789, started his astronomical career as a doorman and porter at the Marseilles Observatory. His interest in astronomy was encouraged by the directors of the observatory, to such good effect that he officially joined the ranks of the astronomers there in 1813, and became Assistant Director of the observatory in 1818, the year in which he pinned down Encke's Comet. He was also one of the people who had caught a glimpse of the comet in 1805. But that was hardly the highlight of his career, which included the directorship of the observatory at Lucca in northern Italy, then that of the one in Florence. Amongst numerous other observations, he was involved in the discovery of 30 out of the 43 'new' comets that were observed in enough detail between 1801 and 1828 to provide reliable orbits; in addition, he found a further seven comets that were observed too briefly for accurate orbits to be calculated. Many of these bear his name, which makes it ironic that the most important of all his comet observations does not. When Encke used Pons' observations to work out the orbit of the comet seen in 1818, he was amazed – as was the entire astronomical community – to discover that it had a period of less than four years. He pointed out that it was the same comet that had been observed in 1786, 1795 and 1805, and predicted its return in May 1822. Encke's Comet[4] duly reappeared, on schedule – the first time that the return of a comet had been successfully predicted since the return of Halley's Comet itself in 1758.

Encke's Comet has since become the most observed comet of them all, as it returns to view regularly. The period of its orbit is decreasing steadily, as gas jetting out from the nucleus during its repeated close encounters with the Sun acts like a rocket, changing its orbit; the period is now about 3.3 years, as it moves in an orbit stretching from 0.34 AU (closer than the average distance of Mercury to the Sun) and 4.08 AU (almost as far out as the closest Jupiter gets to the Sun). But the brightness of the comet (or rather, its dimness) has stayed much the same over two centuries of observation.

Although the orbit of Encke's Comet crosses the orbit of the Earth every three years or so (it is, as Clube and Napier point out, the sole example of an active comet in an Apollo-type orbit), at present the comet misses the Earth by a wide margin every time it crosses our orbit. But there is a scattering of cosmic debris, associated with the comet, that stretches right around its orbit, so that the Earth ploughs into it twice a year, once on each side of its orbit, as it moves around the Sun. This produces showers of meteors ('shooting stars') at the same times each year.

It was only in the 1860s that astronomers made the connection between meteors and comet debris. They already knew that there were regular showers of meteors at certain times of the year, each seeming to come from a particular point on the sky. The meteor showers are identified by the name of the constellation in which they seem to come from is located. The Leonids, for example, arrive every year on 17 November (or a day either side) and seem to come from a point in the constellation Leo. This is a perspective effect caused by a combination of the way the Earth is orbiting through space and the way the stream of particles that make the meteor shower ('meteoroids') is moving through space.

The Leonids occur every year, but in 1833 they produced a particularly spectacular display, which roused the interest of astronomers, who knew that similar spectacular Leonid showers had occurred in 1799 and 1698. The American mathematician Hubert Anson Newton identified 13 occasions in history when particularly impressive Leonid showers had been observed, going back to the year 902. He inferred, in a paper published in 1863, that the cosmic debris that makes up the shower must be in an orbit round the Sun that takes 33.3 years to complete, with a concentration of debris at one point in the orbit and a scattering of cosmic dust trailing around the whole orbit. If these ideas were right, the Leonids should put on another spectacular display in 1866 – and they did.[5] Even better in 1865 astronomers observed a comet, known as Tempel–Tuttle, which also has a period of just over 33 years, moving in the same orbit around the Sun as the Leonid meteoroid stream. Giovanni Schiaparelli also calculated the orbit of the meteoroid stream, and made the connection with the comet in the mid 1860s; he also pointed out that the orbit of another comet, Swift–Tuttle (identified in 1862) was just right to account for another meteor shower, the Perseids, which arrives on or around 11–12 August each year.

One unfortunate consequence of the discovery of the link between comets and meteor showers was that, since meteor showers are produced by pieces of cosmic grit, for nearly a hundred years many astronomers assumed that comets were some kind of flying gravel bank – conglomerations of gritty particles moving together through space, with the outermost particles drifting away from the main

137

bunch and trailing behind. It was only when Fred Whipple came up with the dirty snowball model that a proper understanding of comets began to emerge.

As it happens, Whipple was also involved in improving astronomers' understanding of the meteors associated with Encke's Comet, which are known as the Taurids. The Taurids do not arrive in a single night, or two, like most meteor showers; instead, they are spread over about 12 nights, from 3 November to 15 November each year (the corresponding showers on the other side of the Earth's orbit, the 'Beta Taurids', arrive between 24 June and 6 July). Among other things, this is a sign that the meteoroids that make up the Taurid stream are ancient debris in an old orbit, which have gradually got spread out by the influence of the gravitational pull of the planets. Within this broad spread of debris, there are two main branches, roughly equal in strength, coming from slightly different parts of the sky.

In 1940, Whipple showed that two of the main Taurid showers come from particles in orbits which are exactly the same shape as that of Encke's Comet, but shifted round slightly from it in space. He proved that this effect is caused by the gravitational pull of Jupiter, which in effect tugs the orbit of Encke's Comet sideways round the Sun, once every 5,800 years. Pieces that break off from the comet nucleus get tugged sideways at slightly different rates; and Whipple pointed out that the amount of spreading observed in the Taurid streams would take at least a thousand years to occur, proving that Encke's Comet had been in essentially its present orbit for at least that long. As Whipple put it in his book *The Mystery of Comets*, Encke's Comet is a 'Methuselah of comets, having survived passages within Mercury's orbit for many hundreds, if not several thousands, of revolutions.' But Clube and Napier take the story a giant step beyond even that conclusion.

They point out that out of some 80 known Apollo asteroids, no less than seven (nearly ten per cent) are associated with the Taurid stream. The largest identified Apollo asteroid, Hephaistos, is about 10 km in diameter, and has an orbit which is equivalent to that of Encke's Comet except that it has been shifted sideways so much that it is now oriented at right angles to the orbit of the comet. If this asteroid did indeed split off from the nucleus of Encke's Comet and gradually got separated from it (and it is hard to see how else it could have got into this precise orbit), it would have taken 20,000 years to reach its present orientation. Another of the Apollo objects associated with Encke's Comet is Oljato, a dark object about 1.5 km in diameter, usually regarded as an asteroid. But Oljato occasionally emits puffs of gas, as if it is, in fact, an almost dead comet. Calculating the orbits backwards in time, Oljato and Encke's Comet were in the same place 9,500 years ago.

Other calculations show that there is less than one chance in a million that three objects would follow the same orbit around the Sun as Encke's Comet by chance. But because astronomers estimate that there are about 20 times as many Apollo asteroids as they have yet seen, there may be 20 times seven, or about a hundred and fifty, pieces of debris each more than a kilometre across associated with the Taurid stream. As Clube and Napier say, in their book *The Cosmic Winter*, 'it seems clear that *we are looking at debris from the break-up of an extremely large object*' (their emphasis).

Backtracking the orbits of several Taurid meteors, Whipple and his colleague S. E. Hamid suggested, as long ago as 1952, that a major break-up of material from Encke's Comet occurred about 2,700 years ago, probably when a large piece of comet nucleus was broken off in a collision with an asteroid in the main asteroid belt, between Mars and Jupiter. And the whole accumulation of material – Encke's Comet itself, the associated Apollo asteroids and meteoroidal debris – fills a tube around the Sun so big that the Earth enters it in April each year, does not emerge until late June, then re-enters the tube on the other side of the Sun in October and re-emerges in December. We actually spend nearly half of each year running through the extended stream of cosmic debris associated with Encke's Comet and the Taurid meteors. The Taurid meteor showers themselves simply represent the densest accumulation of cosmic grit in the middle of the tube.

Clube and Napier estimate that the total mass of the material in the tube is about ten thousand billion tonnes, and that it all comes from the break-up of an object that was originally about 100 km across, just the same sort of size as Chiron or QB1. This object must have entered the inner Solar System at least several tens of thousands of years ago, and suffered a major collision, or series of collisions, with other objects that split Encke's Comet and Oljato from the parent body. (It may also, of course, have been partly broken up in the first place either by tidal forces, or outgassing, or both.) If this theory is correct, there are hundreds of thousands of objects in the tube each large enough to cause an explosion equivalent to several megatonnes of TNT if they collide with the Earth – and the Tunguska event itself, occuring close to the time of the Beta Taurids, was probably caused by one of them.

It's worth emphasising that the scenario expounded above is grounded quite solidly in factual observations and real probabilities. We know that objects like Chiron and QB1 exist. Computer simulations show that some of these objects must fall in to the inner Solar System and break up. And we see all around us in our part of the Solar System debris from the break-up of such an object. Astronomers can even calculate, from their present orbits, just when some of these pieces broke apart from one another. Some of the fragments have even struck close to home, quite recently. When one of the larger objects in the stream breaks up, for whatever reason, it will produce a group of meteoroids travelling together through space, not unlike the stream of 'machine-gun bullets' produced by the break-up of Shoemaker–Levy. Clube and Napier point out that between 22 and 26 June 1975, right at the time of the Beta Taurids, seismometers left on the Moon by the Apollo astronauts recorded the impact of a swarm of boulders, each weighing about a tonne, with the lunar surface. As many impacts that size occured in five days as had been monitored over the entire previous five years.

And the Taurid material isn't the only evidence of cometary break-up in the inner Solar System. There is a whole family of comets that pass very close to the Sun (called 'sungrazers'), known as the Kreutz family (after the German astronomer Heinrich Kreutz, who studied them in the 1880s). Orbital calculations carried out by Brian Marsden at the Smithsonian Astrophysical Observatory in the 1960s show that they formed from the break-up of a single very large comet that was seen by Aristotle in 371 BC.

It seems that there is a real and direct link with what is going on in the inner Solar System today and what went on at the time of the death of the dinosaurs. The difference is that this time the Earth has not (yet) been struck by a really large piece of cosmic debris. Even so, the situation is alarming enough that, as we shall see later, serious efforts are now being made to provide some warning of cosmic impacts, and to do something about preventing them, if possible. But there is another threat to the terrestrial environment, more insidious than the possibility of a major cosmic impact – a threat which is also associated with the streams of debris produced by the break-up of supercomets.

Large lumps of rock or ice are the more visible threats among this debris. But a great deal of the mass of the original supercomet ends up in the form of a fine ring of dust around the Sun. There is just such a ring of dust today. It reflects light from the Sun to produce the so-called zodiacal light, visible, most clearly from the tropics, just after sunset or just before dawn. The zodiacal light may be pretty, but it is as threatening, once it is understood, as a lump of rock heading our way. For such clouds of dust could, some astronomers believe, block out enough sunlight when they first form to trigger the onset of Ice Ages on Earth.

Notes

1 Chiron's closest approach to the Sun, at 8.5 AU, occurred on 14 February 1996; it was closest to the Earth on 1 April 1996, a bonus for astronomers, though still, of course, further away from us than Jupiter. It will not come so close again until 2047.
2 There is an exception to this rule. Although Pluto is tiny compared with the giant planets, and its orbit crosses the orbit of Neptune,

the period of Pluto's orbit is exactly 50 per cent longer than the time it takes Neptune to orbit once around the Sun, so that it is gravitationally locked in what is called a 'resonance' with Neptune; which means it will keep its orbit stable for the same sort of timescale that Neptune's own orbit is stable.

3 Chaos also affects the inner planets of the Solar System. Computer simulations show that under the gravitational influence of the giant outer planets, Mercury, for example, can drift into an elongated orbit from which a close encounter with Venus could send it careering unpredictably across the Solar System, and Mars could also be disturbed on timescales of many billions of years. But these are not problems to worry about on the timescale of the life of a human being, or even of a human civilization.

4 It's worth pointing out that Encke himself was embarrassed by having his name attached to someone else's discovery, and tried to get the comet named after Pons, but to no avail.

5 The orbital period is just over 33 years, so the latest spectacular Leonid display was in 1966, when people in the US were treated to the sight of the sky raining meteors at a rate of up to 40 per second. Another comparable display is anticipated in 1999. But these showers of cosmic fireworks pose no direct threat to us, since they are caused by particles, no bigger than grains of sand, burning up in the atmosphere of the Earth.

VII

COSMIC WINTERS

Ice Ages have fascinated scientists and non-scientists alike for nearly two hundred years. The Scot James Hutton first suggested that the scars still visible today on the Jura Mountains of France and Switzerland were created by glacial activity, and the Swiss Louis Agassiz took up the idea and promoted it vigorously. The story of how an understanding of Ice Ages developed in the 19th century is not unlike the story of how geologists in recent decades eventually persuaded themselves of the reality of the impact event(s) at the end of the Cretaceous. Painstaking field studies brought in evidence that was pieced together by theorists into a coherent picture, building up in the face of strong opposition from some quarters incontrovertible proof that great sheets of ice had once scoured Europe and North America. But was it *only* once? Had there been just one Ice Age, or many?

It is now clear that the Earth experiences repeated refrigerations, on two very different sorts of timescales. The first pattern corresponds to large-scale changes in the geography of the globe, caused by the way the continents drift slowly around the surface of our planet. This was itself a controversial idea as recently as the 1950s, but became well-established in the 1960s as various pieces of geological evidence pointed to the same conclusion. We do not need to go over the geological arguments here, not least because it is now possible to measure continental drift directly, using laser beams bounced off satellites orbiting the Earth. These measurements show, for example, that the Atlantic Ocean is getting wider at a rate of a couple of centimetres a year.

Geophysicists are able to reconstruct the way the continents have moved over many millions of years, and can produce maps showing the geography of the globe at different epochs in the past. They can also compare these maps with geological evidence of the kind of fauna and flora that flourished in different parts of the world at those times, and infer what the climate was like on those ancient continents. The key feature of these reconstructions, as far as our present story is concerned, is that they show that for most of its history the Earth has been much warmer than it is today. Why? Essentially because throughout most of geological history there have been no land masses blocking the access of warm ocean water from the tropics to the poles.

Today, there are ice sheets covering both poles of our planet, and the shiny surface of the ice (or the snow lying on the ice) reflects incoming solar heat away into space, chilling the regions around the poles. But if there were no continents in the way, warm water could push up from the tropics all the way to the North and South Poles. When warm water reaches the polar regions, it keeps them free from ice, and this has a major effect on the climate of the whole world: instead of white sheets of snow and ice reflecting heat away, the poles are covered by dark water which absorbs heat from the Sun.

Since only a third of the surface of the planet is covered by continents and since they drift around the globe, most of the time there is no land over either pole. Occasionally – at intervals of hundreds of millions of years – a large land mass drifts across one pole, plunging one hemisphere into a long period of cold called an Ice Epoch. Even more occasionally, there will be ice caps over both poles, as there are today. This is such a rare situation that the exact pattern we have today may even be unique in Earth history. The unusual feature today is that there is land (Antarctica) covering the South Pole, where huge glaciers and ice sheets have built up, while in the north there is a polar sea (the Arctic Ocean) – but it is almost entirely surrounded by land, making it difficult for warm water from the south to penetrate the polar sea, and allowing a thin skin of ice to form. Even a thin skin of ice is very efficient at reflecting away incoming solar heat.[1]

Over the past few million years, the Earth has experienced a 'double whammy' of ice, with both poles affected, keeping it in a prolonged and very widespread Ice Epoch. As we discussed in our book *Being Human*, it is probably not a coincidence that this period of unusual cold has seen the emergence of *Homo sapiens* and the rise of civilization: because during Ice Ages the region of Africa where our ancestors emerged is subject to repeated droughts, which put a premium on the kind of survival skills that go hand in hand with intelligence.

This brings us to the other kind of Ice Age pattern the Earth experiences: changes that take place *within* an Ice Epoch. On a timescale of a few million years, we can largely ignore continental drift, and although the present Ice Epoch represents a cold spell compared with the entire history of the Earth, to us it is normal. But even within the

Ice Epoch, there are substantial variations. These have relatively little effect on the southern hemisphere, where Antarctica is always covered by much the same amount of ice, and it is hard for any extra ice to spread out over the surrounding ocean. In the northern hemisphere, things are more delicately balanced, and there is plenty of land near the polar regions for snow to fall on and build up ice sheets if the region cools. Sometimes the ice surges down from the north across much of Europe and North America, while the heartlands of our ancestral African home experience drought; this is called an Ice Age proper. At other times, the ice retreats to more or less its present position, and the African drought eases; this is called an interglacial. We are living in an interglacial, a relatively warm spell of about 10,000–15,000 years during an extended Ice Epoch.

Geologists have identified a distinctive rhythm to the advance and retreat of ice sheets during the present Ice Epoch, and astronomers and climatologists think they know the cause. The saga of how this pattern was unveiled, and the struggle of theorists to convince doubting geologists that it was real, has been described most lucidly by John Imbrie and Katherine Palmer Imbrie in their book *Ice Ages: Solving the Mystery*. The key to the pattern is that the way the Earth tilts and wobbles as it orbits around the Sun (like a child's spinning top wobbling as it spins) changes the amount of heat arriving at different parts of the world at different times of year. The total amount of heat received over the whole year is always the same, but the balance between the seasons alters.

When the northern hemisphere has cool summers, this allows some of the snow that fell the previous winter to stay on the ground near the Arctic ice cap all year, reflecting away heat and cooling the region still more. So cool summers are associated with the onset of Ice Ages.[2]

But when the orbital variations conspire to bring intense summer warmth to the northern hemisphere, the ice sheets on the land surrounding the Arctic Ocean melt away in the heat. So hot summers are associated with the onset of interglacial periods.

Overall, the orbital changes combine to produce a pattern in which Ice Ages roughly 100,000 years long are separated by interglacials roughly 15,000 years long, a pattern which has persisted for several million years – the Ice Epoch. The 'normal' state of the northern hemisphere is a full Ice Age, and short-lived interglacials only occur when all of the orbital effects pull together to produce maximum summer warmth. The whole pattern is known as the Milankovitch Model of Ice Ages, after the Serbian astronomer Milutin Milankovitch, who was one of the key players in developing the idea.

There is no doubt that the astronomical rhythms of the Milankovitch Model have modulated the climatic state of the Earth during the present Ice Epoch. But that isn't quite the whole story. Although these cycles are very good at producing ripples of warming and cooling within the Ice Epoch, they cannot on their own explain exactly how the world got into its present state, and in particular they cannot explain what caused the Arctic Ocean to freeze.

The point is that the Arctic Ocean today – given the present arrangement of the continents and pattern of solar heating – could quite happily exist either with an ice cap, as it has now, or without. This is, in fact, a rather simple example of chaos at work. If the Arctic region warmed by a few degrees Celsius, the ice cover would melt, exposing the dark ocean water below. Because the dark water absorbs heat that ice reflects away, it would then get warmer still, plunging the world into a very lopsided climatic state, with the South Pole still covered by great ice sheets. But if the ice-free Arctic Ocean were then cooled back to the temperature it is today, the ice sheet would *not* re-form! Because dark ocean water is so good at absorbing solar heat, the surface of the sea would stay warm enough to prevent the ice forming. You would have to cool the globe by several degrees more before the ice suddenly began to spread across the Arctic Ocean again.

There is nothing mysterious about this behaviour. Like a supercomet in an orbit from which it might be ejected out of the Solar System or might be sent plunging in close to the Sun, the Arctic Ocean can be pushed in one of two directions by external factors. Today, the relevant external factor is the rise in temperature caused by the build-up of greenhouse gases produced by human activities.[3] Many people around today may live to see an ice-free Arctic, with all that that implies. But, equally, the Arctic only got to be ice-covered because something gave it a push in that direction, two or three million years ago when the present ice epoch was young. Other things being equal, even with its surrounding fringe of continents the Arctic would never have frozen. Clearly, other things were not equal. Something happened to plunge the northern hemisphere into deep freeze. And that something could well have been the break-up of a giant cometary nucleus in the inner Solar System.

The first person to develop a fully worked out scientific theory along these lines was Fred Hoyle, initially in a series of scientific papers with his colleague Chandra Wickramasinghe, and then in his book *Ice*. He pointed out that the climatic balance of the Earth today rests on a knife edge, and that only a relatively modest tilt of that balance is needed to change dramatically the way in which the 'heat engine' that drives the circulation of the atmosphere works.

The key component of that heat engine is water vapour, which is produced when solar energy is absorbed by the sea, and which gives up that energy when it condenses once again to form rain or snow. If you leave a pan of water boiling on a stove until all the water has evaporated, the heat from the stove has gone into the water molecules, changing them from liquid to vapour. And when those water molecules condense, changing back from vapour to liquid water, they give up the same amount of heat. Water vapour carried from the tropics to higher latitudes carries heat energy along with it, in the form of so-called latent heat. In addition, water vapour in the air is very effective at trapping heat through the greenhouse effect, and keeping the surface of the Earth warmer than it would otherwise be. Water vapour is carried high into the atmosphere, to the cloud tops (at the top of the troposphere, 10 to 15 km above the surface of the Earth). There, the temperature is as low as −20 °C, but the water vapour doesn't all turn to ice, because it can only form ice crystals where there are tiny particles of dust or other material in the air to act as 'seeds'. Without these 'cloud condensation nuclei', it stays as supercooled water vapour even at these low temperatures. It is only when the temperature falls below −40 °C that water vapour changes spontaneously into a profusion of tiny ice crystals, which then act as the condensation nuclei for more water vapour to freeze, forming a mass of particles so reflective that they are known to Antarctic explorers as 'diamond dust'.

This form of ice is so reflective, indeed, that if you took a layer of water equivalent to a thickness of one hundredth of a millimetre over the entire surface of the Earth and turned it into diamond dust crystals, each about one millionth of a metre across, it would reflect almost all of the incoming solar energy back into space. Hoyle pointed out that turning just one-tenth of one per cent of the water vapour in the Earth's atmosphere into diamond dust would have a catastrophic effect on climate. Continents would freeze immediately beneath the blanket of ice crystals high in the atmosphere. For several years huge storms would rage as the warm surface layers of the seas continued to give up water vapour to the air, where it would be carried by the circulation of the atmosphere over the cold land and fall as snow. In a few years, the storms would subside, and calm would be restored, as the stored heat in the upper layer of the oceans was dissipated. Then, even if the diamond dust did disappear from the upper atmosphere, the Earth below would stay in its frozen state; with no reserves of heat in the oceans and the continents covered in a white blanket that would continue to reflect away the incoming solar heat.

Such a disaster could be triggered by reducing the amount of rainfall around the world. As we have mentioned, when water vapour condenses it gives up latent heat to its surroundings, and it is this latent heat that keeps the upper atmosphere as warm as -20 °C. Even though most of the vapour cannot condense, because there are no nuclei for it to condense on to, the small proportion that does condense gives up enough heat to stop the temperature falling further. In fact, each gram of water vapour that condenses into liquid form at 0 °C releases 600 calories of energy – enough energy to raise the temperature of *six* grams of water at 0 °C all the way to boiling point. If the water vapour cools all the way to ice, it releases even more energy – an additional 80 calories per gram. Hoyle calculates that the amount of heat needed to keep the upper atmosphere too warm for diamond dust to form is equivalent to having an average precipitation equivalent to about 65 cm of rain a year (just over 25 inches per year). The only regions of the globe where this does not happen today are around the poles, where diamond dust does indeed form in the high atmosphere, and helps to maintain the polar chill by reflecting away some of the incoming sunlight.

If the oceans were cooled, so that less water vapour gets in to the air in the first place, this would reduce precipitation worldwide, and thereby encourage the formation of high-altitude diamond dust. After all, without water vapour there will be no rain. But there will always be enough water vapour to make diamond dust – remember that you only need one tenth of one per cent of the present amount if the temperature falls far enough, high above our heads. The warmth of the upper layer of the ocean today is equivalent to about a ten-year store of solar energy, so the oceans would have to be cooled for at least ten years in order to tip the Earth into deep freeze. A layer of fine particles suspended high in the stratosphere would do the job very well, by blocking out some of the incoming solar heat.

This 'dust veil' effect has been seen at work many times after major volcanic eruptions, including the recent eruption of Mount Pinatubo, in the Philippines. This threw so much finely dispersed material into the stratosphere (the layer of the atmosphere immediately above the troposphere) that the entire world cooled by about half a degree Celsius for a couple of years. It is hard to imagine a super-volcano that would cool the Earth sufficiently, and for long enough, to trigger the formation of diamond dust on the scale Hoyle suggests: but the break-up of a super-comet in the inner part of the Solar System could provide plenty of dust to do the job.

This is where we part company from Hoyle, at least partly. His own development of the diamond dust theme envisaged meteorite impacts as the source of the dust in the air, a precursor of the nuclear winter scenario. In a further complication, Hoyle actually suggested not only that every Ice Age within the Ice Epoch is triggered by meteoritic impact, but that meteorites are also needed to bring the *end* of each Ice Age and initiate an interglacial. His argument was that stony meteorites (which outnumber metallic meteorites by about ten to one) would send dust high in the air and cool the Earth; while metallic meteorites would spew metallic grains high into the air, where they would absorb solar energy and raise the temperature above the point where diamond dust could persist. All this seems unnecessarily complicated, and it is very hard to see how, even with ten stony meteorites for every metallic meteorite, a random succession of impacts could produce the regular variations from Ice Age to interglacial which have been characteristic of the past couple of million years, and which match so well the predictions of the Milankovitch Model. We think that the introduction of metallic meteorites into the story is a red herring, and that while the Milankovitch mechanism may not be adequate to explain how Ice Ages begin, it is certainly adequate to explain how they end, as we shall discuss shortly.

The great strength of Hoyle's hypothesis is that it explains how the Earth, and in particular the Arctic region, could have been tipped into a frozen state when the latest Ice Epoch began. All we need is enough cometary dust dumped in the region of the Solar System that the Earth's orbit passes through. How much is enough? And could the break-up of a single supercomet supply it? The best way to tackle those questions is through investigations of the dust that exists in our part of the Solar System today, and which produces the phenomenon known as the zodiacal light.

The zodiacal light is one of those sky phenomena that were familiar to our ancestors, but which modern city-dwelling, TV-watching folk may know nothing about. In order to see it clearly, you need to be far away from city lights, on a night with little or no moonlight, watching either the western sky just after sunset or the eastern sky just before dawn. What you will see is a pyramid of light pointing upward from the horizon. The light is reflected sunlight, scattered from huge numbers of tiny particles in orbit in a ring around the Sun. It is often so bright that someone who doesn't know what it is may think it is the direct light from the Sun, the twilight of sunset or sunrise itself – indeed, in the *Rubáiyát of Omar Khayyám*, written around AD 1100, the zodiacal light is referred to as the 'false dawn'.

In 1854, the German astronomer T. J. Brorsen noticed another, fainter patch of light on the night sky exactly opposite the position of the Sun, and named it the *gegenschein*, or 'counterglow'. It is produced by light from the Sun being reflected back towards us by particles in the same ring of dust that makes the zodiacal light. And Brorsen himself later found an even fainter band of light linking the zodiacal light and the *gegenschein* in a strip right across the sky, close to the plane of the ecliptic in which the planets move around the Sun.

The dust which produces the zodiacal light – called, logically enough, zodiacal dust – is largely cometary debris, although some of it may come from the asteroid belt. At present, the amount of zodiacal dust is fairly modest, amounting to about a hundred billion tonnes, roughly the mass of a single cometary nucleus 3 km in diameter. But each grain of dust may be only about 10 millionths of a metre (10 microns) across, and altogether there are estimated to be one thousand billion billion billion (10^{24}) of these particles in the ring of zodiacal dust today.

This ring of dust around the Sun covers much of the inner Solar System, and embraces the orbit of the Earth around the Sun, so that our planet is constantly ploughing through a cloud of fine particles in space. Particles from the zodiacal dust band have been found both in the upper atmosphere, where they can be swept up by collectors (literally, sticky surfaces covered in grease) mounted on high-flying aircraft such as NASA's ER–2 research plane; and also in near space, where they are captured in similar dust traps carried into orbit on artificial satellites and returned to Earth for analysis.

Altogether, several thousand tonnes of zodiacal dust enter the Earth's atmosphere every day, and most of it falls gently to the ground or sea, since the grains are too light and fluffy to burn up as meteorites. But this isn't the only way dust gets lost from the ring. Each particle spirals gently in towards the Sun, and the Sun itself swallows up about 300 million tonnes of the stuff each year – roughly 10 tonnes every second, and enough to eat up the entire ring in a hundred thousand years, if it were not renewed in some way. The ring of zodiacal dust is a dynamic entity, repeatedly being fed with new material from the break-up of comets and constantly losing material, mainly to the Sun. The cometary input, though, is not even, and there must be times, in the aftermath of a comet break-up, when there is much more zodiacal dust than we see today.

Indeed, there may well have been more zodiacal dust than there is today in historical times. Since the present mass of dust is less than the amount you would get by breaking up the nucleus of Halley's Comet, then any time even a modest comet disintegrates near the Sun there will be a distinct increase in the amount of dust and therefore in the brightness of the zodiacal light. Victor Clube and Bill Napier have drawn attention to ancient references, from the early Egyptian, Babylonian and Persian civilizations, that seem to refer to a zodiacal light brighter than the Milky Way. They also point to ancient Egyptian descriptions of the Sun as a 'winged disc' – an indication that only a few thousand years ago there was so much zodiacal dust that, even if the 'wings' were not visible in full daylight, the connection between the Sun and the zodiacal light at sunset and sunrise was obvious. The source of the injection of dust into the zodiacal ring that produced these spectacular displays may have been the break-up involving Oljato and Encke's Comet, mentioned in the previous chapter, that happened around 9,500 years ago.

But the trigger for an Ice Age would have involved a much more spectacular event, the break-up of a Chiron-sized object in the inner Solar System. If an object like this, with a diameter of about 200 km, switched into an orbit like that of Encke's Comet, then even without breaking up completely it would lose dust at a rate of a thousand billion tonnes a year, simply by outgassing. Adding this to the zodiacal cloud of material in orbit around the Sun, and allowing for the rate at which material is being lost from the cloud, it turns out that during the lifetime of such an object in its Encke-like orbit (which might be several thousand years), the zodiacal cloud will contain 300 times as much matter as it does today. Out of this enhanced zodiacal cloud of dust, about six or seven hundred million tonnes of material will be swept up by the Earth in its orbit each year.

The tiny dust particles gathered in this way settle only slowly through the Earth's atmosphere – taking as long as a year to reach the surface of the planet – so on average the burden of cosmic dust in the air is equal to one year's supply. On these figures, the veiling effect of this dust layer would be enough to reduce the amount of sunlight reaching the surface by at least one or two per cent. It is a useful rule of thumb that each one per cent decrease in solar heating will produce an average cooling of about 1–2 °C at the surface of the Earth. So the immediate influence of such an enhanced dust veil would be to cool the Earth by a degree or so. That may sound modest, but it actually amounts to as much as the largest fluctuations in climate that have occurred over the past thousand years. If it happened today, it would correspond to setting the climate back from mid-twentieth century conditions to the chill of the seventeenth century, a period known as the 'Little Ice Age'.

On a timescale of decades, this cooling would not be enough to trigger a full Ice Age. But it is possible that over a few thousand years the oceans would lose enough of their stored heat to trigger the production of diamond dust in the upper atmosphere.[4] There is, though, no need to depend on this slow process tipping the balance: at some point the giant cometary nucleus is sure to break up into pieces, releasing much more dust and also increasing the probability that one or more of the larger chunks of cometary material will strike the Earth. It is also possible that the cooling caused by the enhanced zodiacal dust itself would be enough to trigger an Ice Age directly: at the upper end of the range of estimates, the cooling could be as much as 4 °C, which takes us halfway back to Ice Age conditions, and is enough to trigger the diamond dust effect.

Moreover, we also have to take into account the effect of all the dust in the zodiacal cloud between the Earth and the Sun, which would diminish the amount of solar heat arriving at the top of the atmosphere, before it had to struggle through the enhanced atmospheric dust veil on its way to the surface of the Earth. That could double the overall cooling effect, plunging the planet straight into an Ice Age even without the help of diamond dust (although the diamond dust would then help to keep the planet cool even after the cosmic dust had cleared).

The probability of any or all of this happening is, according to Clube and his colleagues, much higher than anybody thought until recently. The chance of an individual object in a Chiron-like orbit being injected into the inner Solar System within 100,000 years is 80 per cent. This means that, with the growing number of known supercomets in the outer Solar System, it is likely that an object more than 100 km across crosses the Earth's orbit every 400 years or so. Many of the spectacular comets recorded throughout history may well have been spectacular because they were so big, and not, as used to be thought, because they came unusually close to the Earth.

But crossing the Earth's orbit is not the same as striking the Earth, of course, and nor does it mean that such objects must be captured into Encke-like orbits. Most of them pass harmlessly through the inner Solar System on their way around the Sun, and back out into the depths of space, after providing their heavenly spectacle. Only about one in a hundred gets captured into a dangerous (for us), close orbit around the Sun. But maybe the rest do not all escape entirely, even if they are not captured into a close orbit around the Sun. Dynamical studies suggest that the rings of Saturn are unstable, and cannot exist in their present form for more than a hundred million years. Are they the product of the break-up of a giant comet?

We may not be lucky enough to see such a spectacular event as the formation of a new ring around one of the giant planets (although Shoemaker–Levy came close!), and we may not be lucky enough to see a giant comet passing through the inner Solar System in our lifetimes. But Mark Bailey calculates that a comet with a nucleus bigger than 100 km across ought to cross inside the orbit of Jupiter (although still staying outside the orbit of Mars) about once in a human lifetime. Many astronomers are eagerly awaiting such an opportunity to investigate such an object at what would be for them, with the aid of their large telescopes, relatively close range.

A giant comet that ended up in an orbit like that of Halley's Comet, rather than an even shorter-period orbit of the Encke's Comet variety, could influence events in the inner part of the Solar System over an interval of one or two million years. This influence would include an overall increase in the amount of dust in the zodiacal cloud, with a corresponding lowering of temperatures on Earth, increased susceptibility to Ice Ages, and a higher risk of repeated impacts of pieces of the supercomet with the inner planets.

On this picture, even impacts a million years apart from one another in the geological record may represent members of the same cometary shower, and those paleontologists who say that the dinosaurs were in decline before the Terminal Cretaceous Event occurred could still be right, without this undermining the impact theory – the Earth may already have been suffering from the effects of a giant comet before the final blow struck. There is certainly evidence of a prolonged extraterrestrial influence around the K–T boundary itself. As well as traces of meteoritic dust, amino acids thought to have come from space have been found in strata laid down a few hundred thousand years before the Chicxlub impact. And over the same sort of timescale there is geological evidence that the southern ocean, at least, may have cooled by about 4 °C.

With one giant comet being flung into an Earth-crossing orbit every 400 or 500 years, and with one in a hundred of these comets settling into a close orbit around the Sun, the chances are that an enhancement of the zodiacal dust cloud sufficient to produce a major cooling of the Earth will happen about once every hundred thousand years – ten times in a million years.[5] Actual impacts, on the scale of the Chicxlub impact, will be very much rarer than this, because they only occur when the fragments meet the Earth in its orbit. But setting aside that possibility for the moment, the coincidence between this timescale and the timescale of temperature oscillations that has characterised the present ice epoch is tantalising.

We should stress that the cosmic dust is only able to have such a profound effect when the geography of the globe is already primed for cold, with at least one of the poles cut off from warm tropical water. So it is no surprise that ice ages have not occurred at 100,000 year intervals throughout geological history. Indeed, as we have seen, the present geographical distribution of the continents may have made the Earth uniquely susceptible to this chill from space over the past couple of million years. As well as providing the trigger which set the terrestrial thermostat lower and produced an ice-covered North Pole at the start of the Ice epoch, under these circumstances comet dust may well have played a part in modulating the climate within the Ice Epoch.

This is not to say that the Milankovitch mechanism has not been at work as well. The evidence is particularly strong that it is the increase in northern hemisphere summer warmth provided by this mechanism, roughly every hundred thousand years, that temporarily pulls the Earth out of a full Ice Age and into an interglacial. But this turns the original idea on its head, and Milankovitch would certainly have been surprised to see the astronomical rhythms invoked to explain why the Earth occasionally warms *out* of an Ice Age, rather than using them to explain why it repeatedly plunges back into an Ice Age. But we think that that is, indeed, the best way to understand the influence of the Milankovitch mechanism on climate. The most recent Ice Age is (hardly surprisingly) the one we know most about. Studies of annual layers of ice, deposited as snow over the past 150,000 years and drilled from the Greenland and Arctic ice caps for analysis today, show a complicated pattern of behaviour which cannot be solely explained by the Milankovitch cycles. In particular, it shows sudden 'spikes' of cold, dips in temperature of several degrees Celsius lasting for hundreds of years, which are very hard to explain – except as a natural consequence of the injection of extra dust into the zodiacal cloud by fragments of a comet that had suffered its original, major break-up when the Ice Age began.

The latest Ice Age began some 115,000 years ago with a spell of severe cold that lasted for several thousand years, but eased off by 108,000 years ago. It 'only' lasted for some 7,000 years, roughly the same as the length of time between ourselves and the pyramid builders of Ancient Egypt. About 95,000 years ago, the Ice Age tried once again to get into its stride, but once again the intense cold eased after a few thousand years, back to more or less its present day state. Then, in the period from about 80,000 years ago to 70,000 years ago the Ice Age really got going. Temperatures plunged at first, recovered a little bit, and then fell again, initiating a spell of full Ice Age conditions that would last for more than 50,000 years.

Even within these tens of millennia of severe cold, there were variations. Many of the 'spikes' we have mentioned occurred during this interval. There was a relatively mild run of centuries about 30,000 years ago, but this was followed by the most severe weather of the entire Ice Age, with the greatest extent of ice in the northern hemisphere (the so-called 'last glacial maximum') occurring as recently as 18,000 years ago. The Milankovitch cycles only began to tug the world into relative warmth again about 15,000 years ago, and the interglacial was not fully established until between 12,000 and 10,000 years ago. Unfortunately, the favourable Milankovitch state does not last long, and has long since passed its peak. Northern summers are growing cooler again (or would be, if it were not for the anthropogenic greenhouse effect), and the world is once more balanced in a state where the slightest hiccup, in the form of an extra input of cosmic dust, could tip it back into a full Ice Age.

Clube and Napier, at least, are in no doubt that the pattern of temperature fluctuations during the latest Ice Age matches the erratic behaviour of a large cometary break-up in the inner Solar System over the past 115,000 years or so. They point out that the intensification of the Ice Age leading up to the last glacial maximum (inexplicable in terms of the Milankovitch Model) fits particularly well with their dating for the break-up of the extremely large object which has since disintegrated further to form the Taurid meteor stream and its associated phenomena. This disintegration must have happened less than 30,000 years ago, or the fragments from it would have got much more spread out in their orbits and would no longer be recognisable as a related stream of cosmic debris in the inner Solar System.

There are two ways in which the investigation of the influence of giant comets on the terrestrial environment can go from here. One is to focus down on even shorter timescales than the Ice Age/interglacial rhythms. How have cosmic dust, and even impacts on the scale of the Tunguska event, affected conditions here on Earth over the past couple of *thousand* years, rather than over the past couple of *million* years? We shall follow this up in chapter nine. But first, we want to look in the other direction, at what these ideas have to tell us about the great catastrophes, like the K–T event, recurring on timescales of tens of millions of years, rather than a couple of million years. At last, we can put the death of the dinosaurs in its proper perspective, before concentrating, in the remainder of this book, on the implications for ourselves.

Notes

1 The thinness of the skin of ice over the Arctic Ocean is, however, a cause for concern, since even a modest global warming caused by a build-up of greenhouse gases could cause it to melt, totally changing the climate of Europe, Russia and North America.

2 Even though cool summers go hand in hand with slightly warmer winters, as long as it is still cold enough for snow to fall in winter this doesn't matter.

3 The greenhouse effect increases the surface warmth of a planet like the Earth because the atmosphere traps heat that would otherwise escape into space. It works like this. Sunlight passes through the atmosphere without being impeded and warms the surface of the planet. The warm surface radiates energy in its turn, but this is in the form of infrared heat, some of which is absorbed by substances such as carbon dioxide in the atmosphere. This warms the lower layers of the atmosphere, which re-radiate infrared energy in all directions, some going back down to the surface. Substances which are good at trapping the infrared heat are called greenhouse gases, and include carbon dioxide and methane.

4 If you 'lose' one per cent of the Sun's heat each year, then in 100 years you have lost the equivalent of a year's supply of heat, and in a thousand years you have lost the equivalent of 10 years' supply of heat, which brings you close to the point where Hoyle calculates that the diamond dust effect will be triggered.

5 We know that a hundred times 500 is only 50,000, not 100,000; but all these numbers are very approximate, and we are trying to err on the side of caution!

VIII

CYCLES OF DOOM

The kind of cosmic influences on Earth that we discussed in the previous chapter can be regarded as part of the normal, routine life of the Solar System, on a timescale of hundreds of thousands, or millions, of years. On average, from time to time there will always be one or two giant comets falling in to the inner Solar System from the great reservoirs of the Kuiper Belt and the Öpik–Oort Cloud, and the disruptions that they cause are as normal as the disruptions caused by, say, hurricanes or avalanches. Not predictable occurrences, and not desirable, either; but something that life just has to get along with.

But what if some influence from outside were to cause a major disruption of the comet cloud? After all, it does extend halfway to the stars, and it doesn't extend further than that because beyond that point the orbits of the comets would be disrupted by external influences, especially the gravity of other stars. So what happens when the Solar System passes relatively close to another star – or, if you like, when another star passes relatively close by us? As you might expect, this must disrupt the outer cloud, sending not just one or two but many comets on orbits which will eventually take them in to our part of the Solar System. It wouldn't be so bad if all those comets were the size of the nucleus of Halley's Comet; but now we know that these incomers might include giants like Chiron. Such disturbances of the comet cloud are inevitable, on geological and astronomical timescales, and the next one can even be predicted. It won't affect us but it is likely to cause havoc on Earth long before as much time has elapsed from now as the interval between ourselves and the death of the dinosaurs.

Soon – by astronomical and geological standards – the Solar System will experience a series of relatively close encounters with passing stars. One of these, involving the double system Alpha Centaurus A/B will disturb the cloud of comets that surrounds the Solar System, sending hundreds of thousands of comets plunging inward on potentially Earth-impacting orbits. The co-mets won't arrive in the inner Solar System for twenty million years or so. But the calculations which point to this potential for future catastrophe help to show just why cosmic impacts have played an important part in the evolution of the Earth (and of life on Earth) over geological time.

At present, the closest star to the Sun is Proxima Centauri, at a distance of 1.295 parsecs (1 parsec is about 3.25 light years). But there are 57 other stars within a distance of 5 parsecs from the Sun. Most of these stars are smaller than the Sun, and will have little gravitational influence on the Solar System even at their closest approach (distances comparable to the present distance of Proxima).

But the Cen A/B system has a combined mass more than twice that of the Sun, enough to disturb the comet cloud when Cen A/B comes within 1 pc of the Sun in about 28,000 years from now.

Robert Matthews, an amateur astronomer based in Oxfordshire, described the implications in the *Quarterly Journal of the Royal Astronomical Society*.[1] In his calculations, Matthews used the estimate that the radius of the Öpik–Oort cloud is just under 0.5 parsecs, and that it contains about 5,000 billion comets. Putting this in another perspective, the distance to the outer edge of the Oort cloud is about 200,000 times the distance from the Sun to the Earth, or 200,000 AU.

For several thousand years Cen A/B will actually be closer to the comets on that side of the Solar System than those comets are to the Sun. Paradoxical though it may seem, the disturbance caused by the gravitational pull of Cen A/B doesn't just tug 'our' comets towards itself, but can tip comets into orbits taking them into the inner Solar System. Matthews calculates that about 200,000 comets will be perturbed sufficiently by this close encounter to fall in to the inner part of the Solar System – but, falling slowly from such a great distance, they will take about 20 million years to reach the vicinity of the Earth. Perhaps even more significantly, they will not all reach the inner part of the Solar System at exactly the same time. Falling at slightly different speeds over such a long interval of time, they will arrive in a spread-out shower, lasting perhaps for millions of years.

The message from Matthews' work, and similar calculations made by other astronomers, is that a disturbance of the Öpik–Oort cloud leads (eventually) to an increased chance of cometary collisions with the inner planets, probably producing repeated impacts on Earth on a timescale of millions of years.

Crucially, this is exactly the pattern that geologists find at the K–T boundary, and at the times of other mass extinctions. All the species involved in such extinctions do not disappear together in one single event, but 'stepwise', over a period of a few million years. One clutch of plants or animals is wiped out in one step, another in the next step, and so on. And where there are traces of impact craters associated with these stepwise mass extinctions (most notably, as Victor Clube was the first to point out, at the K–T boundary itself, which is the one that has been most intensively studied), the craters themselves are spread out over the same sort of interval.

In the 1980s, geologists and paleontologists struggled to come to terms with the idea that pulses of comets arriving in the inner Solar System had repeatedly extinguished many varieties of life on Earth, and had thereby affected the course of evolution. But the hardest thing they had to come to terms with (indeed, many of them still haven't come to terms with it), was the discovery that these comet pulses may not have arrived entirely at random during the course of geological time. Some of the pulses do occur at random; but the greatest extinctions seem to recur with a rhythm 26 million years or 30 million years long.

The idea that there might be a rhythm of this kind in the mass extinctions of life on Earth actually predates both the realisation that the terminal Cretaceous event was associated with an impact from space and the realisation that the Chicxlub impact was probably just the largest of several impacts that contributed to the end of the era of the dinosaurs. So when, in 1977, Alfred Fischer and Michael Arthur, of Princeton University, published a paper claiming that the large extinctions that had oc-curred over the past 250 Myr were spaced at more or less even intervals of 32 Myr, nobody took the notion seriously. Cycles like this are anathema to most scientists, partly because the human eye and brain have a tendency to 'see' patterns that are not really there when random, 'noisy' data are plotted out graphically. Nobody likes to accept the reality of a cycle unless there is some known physical process to account for it. In 1977 the best 'explanation' that Fischer and Arthur could come up with was that some unknown mechanism associated with convection inside the Earth could drive the extinction process by periodically triggering outbursts of volcanic and earthquake activity.

Without a satisfactory physical mechanism to explain the cycle, the only other reason to take it seriously would be if there was an overwhelming weight of statistical evidence in its favour. That the Princeton researchers simply did not have. Their evidence was suggestive, rather than compelling. And with only eight 32 Myr intervals involved in an timespan of 250 Myr, it wasn't obvious that the statistics could be made compelling. But over at Harvard University a young student, Jack Sepkoski, had already started to compile a database that would eventually provide those reliable statistics even before Fischer and Arthur went public with their idea.

Sepkoski was a graduate student at Harvard from 1970 to 1974 and was encouraged by Stephen Jay Gould, his supervisor at Harvard, and David Raup to compile data on changes in the patterns of life on Earth – data which could be used to study the way life forms had diversified, with new varieties developing in different directions from the same ancestral species. Inevitably, the database included information about extinctions, but that was not its primary *raison d'etre*. Its compilation was an enormous task, which continued for years after Sepkoski left Harvard (he first went to the University of Rochester, and later became Professor of Paleontology at the University of Chicago, which is where David Raup also works). Indeed, the compilation of the database continues today, and will never end, since Sepkoski is always adding new bits of information to it, and correcting minor errors.

He doesn't get the information by going out into the field and digging up fossils himself. Instead, he digs up data from published information about fossils going back to the middle of the 18th century, collating information and putting each fossil in its proper place in the geological record and in the family tree of life on Earth. Like Topsy, the project just growed, but it never seemed to be in a fit state for publication. Sepkoski was always unhappy with the state of the database, sure that there were mistakes in it that still needed correction, as he hauled it off with him from Rochester to Chicago in 1978. It might never have been published if it hadn't been for a misunderstanding that Sepkoski recounts in his entertaining contribution to the book *The Mass-Extinction Debates* (edited by William Glen).

In the spring of 1980, one of Sepkoski's colleagues at Chicago told him that he was coming up for a tenure review. This meant that his job was on the line – if his work was deemed as coming up to scratch, he could stay relatively securely at the university, but if not, he would be out on his ear. It later turned out that the information was wrong, and Sepkoski's position was not on the line. But by then he had already been panicked into action. Searching for something to publish which might impress his peers sufficiently for him to pass the review, he decided to knock the database into shape and offer it to the world. By the time he found out about the mistake concerning the tenure review, he had invested so much effort in this that he went ahead anyway, and the first version of the database duly appeared in print in 1982 as *A Compendium of Fossil Marine Families*.

At first sight, the *Compendium* is one of the most boring volumes ever published, simply a list of names of biological families, together with the places of their first and last-known occurrences in the geological record (essentially, the dates during which each family was alive) and details of the sources of the information. But even before the volume had been published, it had proved its worth.

Raup had received an early copy of the *Compendium*, and used the data to plot a graph showing the rate at which extinctions had occurred over geological time. To the surprise of Sepkoski, who had not realised just how much detailed information the database contained, the graph clearly showed the five biggest extinctions of life on Earth during the past 500 million years or so. This demonstration of the strength of the database encouraged Sepkoski to look at smaller extinction events, not just the five most spectacular ones.

Raup, who described the database as 'like a new toy'[2], was Sepkoski's enthusiastic collaborator in this work. Both of them knew about the study by Fischer and Arthur – indeed, Sepkoski had just published a paper concluding that their claim of a 32 Myr cycle in the fossil record was wrong. But it seems that it never occurred to either of them to use the database to search for periodicities, presumably because their minds were so firmly set against the possibility that such periodicities could exist. So when, in the winter/spring of 1983, Raup came up with a graph showing a distinct hint of a cyclic variation in extinctions

173

of life on Earth, Sepkoski's initial reaction was, 'Oh shit, Fischer and Arthur were right.'[3]

But that was just the beginning. Unlike Fischer and Arthur, Raup and Sepkoski had enough data to carry out a painstaking statistical analysis, coming up with very strong evidence for a repeating pattern of extinctions with a rhythm 26 Myr long. The difference between their 26 Myr cycle and the 32 Myr cycle of Fischer and Arthur is not something to worry about. Different versions of the geological record (different interpretations of the same data) give slightly different dates, and there is always some uncertainty (more uncertainty the further back in time we probe) in all of these dates. If Fischer and Arthur had used the same version of the geological timescale as Raup and Sepkoski, they too would have come up with an interval of 26 Myr between extinctions but in view of the uncertainties, in round numbers you can think of all these cycles as having a rhythm which repeats roughly every 30 Myr. What matters is that, whatever their exact length, the cycles are real. Raup and Sepkoski showed, using detailed statistical techniques, that the pattern found in the record of life on Earth could only occur by chance once in a thousand times – in statistical jargon, the discovery of the cycle is reliable at the 99.9 per cent confidence level. But that still leaves the puzzle of what causes this regular repetition in the cycle of death.

By May 1983, when Raup presented this discovery to a scientific meeting in Berlin, people were already being prepared mentally to consider the possibility that the cause might be extraterrestrial. Raup's presentation to that meeting was deliberately low-key (he wanted to get the discovery of the cycle on record, but without making too much fuss until he had some idea what drove the cycle). The main topic of conversation was the still new idea that the traces of iridium found by the Alvarez team showed that there had been a major impact on Earth at the end of the Cretaceous. Later that year, Sepkoski talked about the 26 Myr cycle at a meeting in Flagstaff, Arizona, which was also convened largely as a result of the interest in extinctions roused by the discovery of the iridium layer at the K–T boundary. This was the first occasion on which either he or Raup suggested that the origin of the cycle might be extraterrestrial. The possibility was only mentioned in the vaguest terms, in the hope that astronomers might take up the unenviable task of trying to find a mechanism to explain the cycle. Those hopes were fulfilled so quickly, and so spectacularly, that the astronomers, with their suggested explanations of the 26 Myr cycle, almost beat Raup and Sepkoski, with their evidence for the 26 Myr cycle, into print.[4]

Feeling no sense of great urgency, Raup and Sepkoski, having tested the waters with their conference presentations, wrote up their discovery and sent it off to the journal *Proceedings of the National Academy of Sciences* in October 1983; it was published in February 1994. But Sepkoski's Flagstaff presentation had been picked up by several science journalists, and was widely reported. Although an early version of the Raup and Sepkoski paper was circulated among their colleagues, their mailing list of regular paleontological and geological correspondents did not include astronomers. So it was through the news stories that astrophysicists first heard of the 26 Myr cycle, and then requested early copies ('preprints') of the paper themselves. Without the news stories – which were, again, published largely because of the interest in extinctions roused by the work of the Alvarez team – the astrophysicists would not have become aware of the 26 Myr cycle until after the *PNAS* paper had been published. As it was, by January 1984, theories relating to the discovery of the cycle were flooding in to the offices of the science journal *Nature*, known as the best place to seek quick publication of exciting new ideas.

One of the people involved in producing that flood of papers, Richard Muller, of the University of California, Berkeley, has told how the claim by Raup and Sepkoski that they had found a 26 Myr cycle of extinctions triggered this wave of activity. His story[5] is not quite typical, because Muller worked with Luis Alvarez, who was, of course, one of the first people to receive a copy of the Raup and Sepkoski preprint. But it is still a neat example of the way science works.

Muller tells how, when he received it late in 1983, Alvarez was really fired up by what he called this 'crazy paper', and drafted an immediate reply attempting to knock down the arguments put forward by Raup and Sepkoski. He asked Muller to check the letter over before mailing it, but clearly expected Muller to agree with his position. So Muller sat down to work through the Raup and Sepkoski preprint; and the more he studied it, the more he became convinced that it was a sound piece of work. This, though, was not what Alvarez wanted to hear. He was sure that major extinctions were associated with impacts from space, and simply could not accept the possibility that asteroids might strike the Earth at regular intervals. In the course of a heated argument with Muller, he insisted that the 26 Myr cycle could not be taken seriously unless there was some physical explanation – a model – of how asteroids could hit the Earth periodically.

In desperation, Muller suggested that there might be a companion star orbiting the Sun – a faint companion star in such a distant orbit that it had never been identified – which only approached the Sun every 26 million years. At the time of closest approach, it might do something to produce a shower of asteroids that struck the Earth. At that moment, Muller was not thinking in terms of comets or disturbances in the Öpik–Oort Cloud, but of asteroids being brought in closer to the Sun by the hypothetical companion star. It was a crazy idea, tossed out without detailed thought. But to his surprise it stopped Alvarez in his tracks. Alvarez thought that the wild idea might work. It was especially attractive to him because a disturbance of the asteroids (or comets) leading to a shower of such objects striking the Earth would fit in with the measurements of iridium and other trace elements at the geological layers corresponding to extinctions. Those measurements showed that the ratios of the amounts of these elements were the same as in other Solar System material, so whatever caused the extinctions had to originate in the Solar System. Comets shaken up by a passing star would do the job nicely.

The next, simple step was to calculate how far from the Sun a faint star in a 26 Myr orbit would be. The answer is 176,000 AU, on average. To cause the kind of repeated disruption the model required, most of the time the star would have to be even further out, dipping down to about 100,000 AU, right through the Öpik–Oort Cloud, every 26 Myr. The model really could be made to work, and within minutes Muller and Alvarez were on the phone to Raup and Sepkoski with the news.

The model is, indeed, even better than anyone realised at the time – because in 1983, remember, the identification of Chiron as an ice dwarf, and the realisation that supercomets exist in the outer Solar System and can be perturbed into orbits crossing that of the Earth, had not been achieved. Still ignorant of that refinement, Muller and his colleagues Marc Davis and Piet Hut worked Muller's wild idea up into a solid scientific theory, which they submitted as a paper to *Nature*. It arrived at *Nature* on 3 January 1984; and in spite of the speed with which they had worked, they were only just in time.

Such a simple model was, inevitably, also hit upon by other researchers at about the same time. And it was not the only astronomical explanation that was almost immediately put forward to explain the 26 myr cycle of death.

Another team intrigued by news of Sepkoski's Flagstaff presentation and the Raup and Sepkoski preprint were Michael Rampino, of New York University, and Richard Stothers, of NASA's Goddard Institute for Space Studies (where Rampino was also a consultant). Their first reaction was to check that the cycle claimed by Raup and Sepkoski was real, by carrying out their own analysis of the geological record. Using a slightly different technique, they found a cycle closer to 30 myr long – which, as we have mentioned, is the same as the Raup and Sepkoski cycle to within the range of uncertainty inherent in all these techniques. Already impressed by the evidence that the K–T event had been caused by an impact from space, they immediately realised that if the same mechanism was at work in other extinctions it ought to show up as a periodicity in the cratering record on the surface of the Earth. Sure enough, when they analysed the dates of known cratering events they found a cycle with a period of about 31 myr. The pieces were beginning to fit together. Now, all they needed was an explanation of why clusters of impact craters should be formed every 30 million years or so.

Rampino and Stothers favoured (and still favour) a variation on the comet shower theme which links the Solar System to the structure of the entire Milky Way Galaxy in which we live. The Milky Way is a typical disc galaxy, consisting of a few hundred billion stars (more or less like the Sun), most of which are in a relatively thin disc of material, about 100,000 light years across, but averaging only about 5,000 light years in thickness (it is thinner still at the edge, and slightly fatter near the centre). There is also a central bulge of stars, giving the whole system something of the appearance of a huge fried egg, cooked easy-over. The Solar System is rather more than halfway out from the centre of the Galaxy, and orbits around the Galaxy roughly once every 250 million years. But as it orbits, it bobs up and down above and below the mid-point of the disc of stars, like the needle of a sewing machine moving up and down through cloth, or, in the analogy Rampino favours, 'like a painted horse on an amusement park carousel'.[6]

Whichever analogy you favour, the relevant point is that it takes – you guessed! – about 30 million years for the Solar system to complete one cycle of bobbing up and down through the plane of the Milky Way. If something happened to shake up the Öpik–Oort Cloud during the crossings of that plane, it would produce comet showers with exactly the right repeating rhythm to explain the pattern found in the extinction record *and* in the cratering record. Rampino and Stothers were, in fact, first off the mark with an explanation of the cyclicity in the extinction record, and their paper along these lines was in the *Nature* office on 15 November 1983.

They beat by just one day a paper by Richard Schwartz and Philip James, which went in to less detail about the cycles but also suggested that the periodicity claimed by Raup and Sepkoski could be caused by the up and down motion of the Solar System through the plane of our Galaxy. Unlike Rampino and Stothers, however, Schwartz and James did not link the cycles of death with cometary impacts. They suggested instead that extraterrestrial influences associated with the passage of the Solar System through the plane of the Galaxy would lead to an increase in cosmic radiation reaching the Earth and to climate changes which caused the extinctions. This doesn't stand up in the light of the weight of evidence that has now accumulated showing that impacts are indeed responsible for the extinctions, and we shall say no more about this model.

There was a lull over Christmas. Then on 3 January 1984 (the same day that they received the Davis, Hut and Muller paper), *Nature* received a paper from Daniel Whitmire and Albert Jackson, also suggesting that the 26 Myr periodicity could be caused by the influence of a companion star to the Sun on the comet cloud. And as icing on the cake, on 30 January 1984 *Nature* received a paper from Walter Alvarez and Muller reporting an analysis of impact crater dates (carried out independently of the work by Rampino and Stothers) which showed a periodicity of 28 Myr.

All five of the papers we have just discussed appeared together in the issue of *Nature* dated 19 April 1984. The Whitmire/Jackson paper actually immediately preceded the one by Davis, Hut and Muller, and offered exactly the same explanation of the cycles of death. But it was Muller's group that made most of the headlines in the resulting wave of media interest in the story, simply because they had actually given the unseen companion star a name – 'Nemesis'. This was tough on Whitmire and Jackson, but less than a year later Whitmire had another stab at the problem, gaining his own fair share of attention when, in January 1985, he published another paper in *Nature* (along with his colleague John Matese) suggesting that the unseen object that was causing the regular comet showers might not be a distant star after all, but an as yet undiscovered planet orbiting beyond Neptune and disturbing the Kuiper Belt, rather than the Öpik–Oort Cloud. Clearly having learned from the events of April 1984, Whitmire and Matese gave this hypothetical planet a name, 'Planet X', which was soon plastered across the headlines.

So by the beginning of 1985 there were three different astronomical explanations for the cycle found by Raup and Sepkoski (four if you include the Schwartz/James variation on the theme), and more evidence than ever before that the cycle, both in the fossil record of life on Earth and in the cratering record, is real. Which model best fits all the facts – which is the 'best buy' now, from a perspective ten years on?

'Planet X' is probably the weakest of the three candidates. It would require a largish planet (Earth-sized, at least, and definitely not just an ice dwarf) in a very strange orbit. Although it seemed just possible that such an undiscovered planet might exist in 1985, after ten years of looking in the right kind of places astronomers have failed to come up with a candidate. In addition, although some minor discrepancies in the orbits of the outer planets seemed in 1985 to offer a hint of the influence of another planet in the outer Solar System, tugging on the known outer planets, improved observations and calculations of the orbits in question over the past ten years have removed the leeway for this speculation. As of now, it seems that the orbits of all the planets can be explained quite satisfactorily without invoking another large planet beyond Neptune – which, turning the argument on its head, means that there can't be a Planet X, or we would see its influence at work on the orbits of other planets.

'Nemesis', on the other hand, is still a viable explanation, although the story of the so-called 'Death Star' has gone off the boil since the mid-1980s, for the simple reason that nobody has yet found a candidate star, and until they do there is really nothing more to add to the story. It might seem surprising that a star so close to the Sun could prove so hard to find, but that is because the word 'star' automatically conjures up an image of a hot, brightly glowing object like the Sun itself. In fact, probably at least half of the stars in the Galaxy are much fainter objects, known as red dwarfs or brown dwarfs.

The names are not just whimsy. The colour of a star (like the colour of a hot, glowing lump of metal) depends on its temperature. Blue-white stars are the hottest, yellow stars like the Sun are in the middle of the range of surface temperatures, and a red dwarf is correspondingly cooler than the Sun. The reason why a star like the Sun shines so brightly is that intense heat is being generated in its interior, by nuclear fusion, to hold it up against the pull of its own gravity – against its weight. When a star begins to form from a cloud of material collapsing under its own weight in space, the interior of the cloud gets hotter as it collapses, because gravitational energy is being released and turned into heat. The more mass the proto-star contains (the heavier it is) the more heat it generates in this way. For a star like the Sun, the collapse halts when the central temperature reaches about 15 million Kelvin (essentially the same as 15 million °C), and nuclear fusion begins. If it were not for the energy generated by fusion, holding the star up, the star would collapse even further and get even hotter inside – which is why astronomers sometimes like to quip that nuclear energy is what keeps a star *cool* inside.

But a proto-star that starts out with much less mass than the Sun never has to generate as much energy to hold itself up. This means that the amount of heat escaping from its surface is less, so it is cooler and fainter. A star with about half the mass of the Sun would be a red dwarf, with very low surface brightness, very difficult to detect even at a distance of only 100,000 AU, even if you knew where to look. And there are probably many even smaller objects in the galaxy, objects with less than a tenth of the Sun's mass, intermediate in size between the Sun and the giant planet Jupiter. Such 'brown dwarfs' never get hot enough inside to trigger nuclear fusion at all, and it is debatable whether they should be called stars. But one of them in the right orbit around the Sun would certainly be capable of causing periodic disturbances to the comet cloud. Nemesis itself need be no brighter than Pluto as seen from Earth (if it is a brown dwarf, it may even be invisible from Earth), and moving very slowly against the background stars of the night sky.

One big item in support of the idea of Nemesis is that rather more than half of all stars are thought to have companions of one kind or another. Identifying a companion for the Sun would actually place it among the more common kind of star. One big item against the idea of

Nemesis is that its hypothetical orbit is so elongated that it is hard to see how the Sun could hold on to its companion for very long, by the standards of astronomical timescales.

Ideas about the orbit have been revised since Muller and Alvarez carried out their quick calculation at the end of 1983. The best guess now is that Nemesis would have to be in a highly elliptical orbit bringing it as close to the Sun as 30,000 AU at perihelion, and taking it as far out as 150,000 AU at aphelion – more than two light years away from the Sun. There is no problem with the resulting influence on comets. During its passage through perihelion, the star should spend about a million years travelling through the more densely populated inner region of the comet cloud, perhaps even disturbing the Kuiper Belt. In their turn, comets disturbed by the passage of the 'Death Star' would take about a million years to fall in to the inner Solar System.

About half the disturbed comets would be deflected inwards, with the other half moving out further from the Sun. The time taken for the ones that come our way is 'only' a million years, because unlike the disturbances caused by the influence of a nearby star they start their journey from the inner part of the Öpik–Oort Cloud, not its periphery. So there would be a continuous rain of comets, rather than a shower, affecting the inner Solar System for about a million years. It would then take Nemesis nearly 13 million years to climb out to aphelion, and 13 million more years to fall back to perihelion, when the whole process would repeat.

Supporters of the Nemesis hypothesis generally interpret the fossil record as indicating that the latest significant extinction of life on Earth caused by the Death Star occurred about 13 Myr ago. This means we are about halfway between such extinctions (fortunately for us) and the star is at its maximum distance from us at present (unfortunately for astronomers), making it extremely hard to detect. The appeal of this chronology is obvious – if the latest extinction caused by Nemesis was 13 Myr ago, the one before that occurred 39 Myr ago, and the one before that 65 Myr ago, smack on the K–T boundary.

Unfortunately, the kind of orbit we have just described is unstable. At aphelion, Nemesis would be almost as strongly influenced, gravitationally speaking, by other nearby stars as by the Sun, and its orbit would be slightly different (perhaps considerably different) every time it passed through perihelion. To the astonishment of the geologists – who are used to hearing complaints that their dates are hopelessly sloppy and inaccurate – the astronomers told them that, as far as the Nemesis hypothesis was concerned, the 26 Myr cycle was 'too good to be true'. For if it were caused by a Death Star the timing of the cycle would shift by at least ten per cent (at random) from one orbit to the next. And in any case, statistical studies (like the statistical studies of the orbits of comets in the Solar System) showed that the Death Star would escape entirely from the Solar System within a couple of hundred million years.

It is still possible that the star does exist, and that either it used to be a separate body wandering through space and was captured by the Sun, or that it used to orbit much closer to the Sun and has relatively recently been disturbed into this unusual, elongated orbit. But either hypothesis requires the unlikely coincidence that we should be around puzzling over the fossil record just at the time when Nemesis has been active – during less than five per cent of the life of the Solar System to date. The only explanation for that coincidence would be that the influence of Nemesis has affected evolution on Earth (not least by wiping out the dinosaurs) in such a way that it has encouraged the emergence of intelligent life forms like ourselves. Although this is an intriguing speculation, and could make the basis of a good science fiction story, it is hard to see how such an influence could really work out in practice.

But the most damning item counting against the Nemesis hypothesis is that there is no reason to expect any companion to the Sun to be in this particular kind of orbit. The orbital parameters are tuned to match the fossil record, and although the orbit that you come up with is a possible one, it isn't a particularly likely one. This weakness in the Nemesis story is the great strength of the rival 'galactic carousel' hypothesis. In that case, the astronomical cycle is already known, and well understood, independently of the cycle in the fossil record. The fact that the two cycles are the same suggests either a bizarre coincidence, or cause and effect at work. So, although the Nemesis idea is by no means dead, and could yet turn out to be the explanation of the cycle of death found in the fossil record by Raup and Sepkoski, our best buy is, indeed, the idea promoted by Rampino and Stothers, that the vertical motion of the Solar System, bobbing up and down through the plane of our galaxy, is the driving force behind the cycle. But how does this motion affect the comet cloud?

In fact, the basis for the mechanism which disturbs comets as the Solar System passes through the plane of the Galaxy had already been put forward in 1982, by Victor Clube and Bill Napier, before anyone realised that there was a cycle of death to explain. They had suggested that catastrophic extinctions of life on Earth might occur when the Solar System encountered a cloud of interstellar material on its journey through space. Going further than most of their colleagues were prepared to go in the 1980s (and further than most astronomers would go even today), they argued that these interstellar clouds are actually the birthplaces of comets, and that during an encounter between the Solar System and an interstellar cloud two things happen. First, the existing Öpik–Oort Cloud is completely disrupted, with most of the comets being swept away into interstellar space, but many of them crashing through the inner part of the Solar System, and some colliding with the Earth. Then, a new halo of comets would be picked up by the Solar System as it moved out of the interstellar cloud.

In this scenario, far from being relics of the formation of the Solar System, the present inhabitants of the Öpik–Oort Cloud are a transient population, plucked from an interstellar cloud, and simply the latest in a series of populations that have inhabited the comet cloud during the long lifetime of the Solar System. If you accept the evidence that extinctions are caused by comet impacts, this is actually a weakness of the Clube/Napier model, because the element abundances associated with those impacts are, as we mentioned, the same as those in the Solar System. It is one extra thing you have to explain (not particularly difficult, but niggling) if those elements were brought in by comets that had themselves evolved quite separately from the rest of the Solar System.

Whatever, the Clube and Napier hypothesis is catastrophism on the grandest scale. They were not, in 1982, seeking to explain the relatively modest extinctions that are, we now know, spaced at intervals of 26 myr to 30 myr through the recent fossil record, but the huge, rare major extinctions that occur every couple of hundred million years or so. Even here there is a hint of a cycle of death at work, with the major extinctions repeating at intervals roughly 200–250 Myr long. This is tantalisingly close to the time it takes the Solar System to orbit once around the Galaxy.

The model actually has echoes of a still older idea, put forward by William McCrea, of the University of Sussex, back in the mid-1970s (and published, of course, in *Nature*[7]. McCrea (who is also one of the few astronomers who believe that comets form in interstellar clouds, and may even be the building blocks of stars) drew attention to the 'coincidence' that the intervals between catastrophic events on Earth, including major Ice Epochs, roughly match the interval between passages of the Solar System through concentrations of material known as 'spiral arms'. These spiral arms twine out from the centre of the Galaxy towards the periphery, and, like every other star in the disc of the Galaxy, the Sun passes through them as it orbits around the galactic centre. The Clube/Napier model would provide one possible explanation of a connection between passages through spiral arms (where the Solar System is more likely to encounter an interstellar cloud or two) and catastrophes on Earth.

There is, indeed, reasonable evidence in the geological record for two cycles of death, one with a rhythm 250 Myr long and the other with a rhythm about 30 Myr long. But we shall only concentrate on the shorter cycle, since the long cycle is so hard to explain that, unless it is simply a larger and more dramatic variation on the same theme that causes the short cycle, we haven't the faintest idea how it originates.

Rampino and Stothers came up with a more modest version of the Clube/Napier idea, suggesting only that during its up and down motion through the plane of the Galaxy the Solar System sometimes encounters interstellar clouds. These interact with the Öpik–Oort Cloud with sufficient strength to disturb some of its component comets into trajectories falling in past the Earth, but not so violently that the entire cloud is stripped away and replaced. In round terms, the time taken for one complete cycle of bobbing is about 62 Myr and, of course, the Solar System passes through the plane of the galaxy twice in each cycle, once going up and once coming down. So the interval between encounters with the denser material of the galactic plane is about 31 Myr, matching the interpretation of the fossil record favoured by Rampino and Stothers.

Supporters of the Nemesis hypothesis (and they still exist!) point to what they regard as a flaw in the Rampino/ Stothers model: the lack, as they see it, of an extinction event in the recent geological past, even though the Solar System is at present quite close to the galactic plane, having passed through the mid-point and now lying about 25 light years above it. Rampino is completely unperturbed by this criticism. There has, he says, indeed been an unusual amount of impact activity in the past few million years. And he points specifically to 'three newly formed craters on Earth, each measuring more than five miles across, and several microspherule layers found in deep sea sediments' which 'suggest that, during the past few million years, there has been an unusual number of comet impacts', supporting the galactic carousel theory.

So why hasn't there been a wave of extinctions in the recent geological past? Actually, there has! Not one of the most dramatic extinctions on record, to be sure, but the latest Ice Epoch, as you might expect, spelled doom for many species of life on Earth. The odd thing is that paleontologists don't always seem to take much notice of this. As far as geological timescales are concerned they can sometimes be rather long-sighted, focusing on more distant events in geological history, and subconsciously regarding recent events as too trivial to bother with (a case, if you like, of failing to see the trees for the wood).

What it boils down to, as far as the two rival astronomical schools of thought are concerned, is whether the recent extinctions should be counted as part of the cycle of death, or whether they are an intermediate, maverick blip with a different cause. For, of course, if there really has been a burst of impact activity in the past few million years, then the previous burst of activity should have been just over 30 Myr ago, and the one before that should have occurred around 65 Myr ago, just at the time of the death of the dinosaurs! *Both* schools of thought are able to make their models tie in with the key event in all of this chronology, the terminal Cretaceous event.

But Rampino can do even better than this. As well as the unusually high level of impact activity in the past few million years, he points out that in the past five million years or so the tectonic activity of the Earth (volcanism, continental drift and so on) has also been unusually high, sea level has been reduced, and the world has experienced an Ice Epoch. All of this matches the integrated model of terrestrial activity that he now links with bursts of impact events. It works like this.

As well as finding evidence that the cratering record shows bursts of activity about 30 Myr apart, Rampino and Stothers find the same rhythm in many other geological/

geophysical processes. These include changes in sea level, changes in volcanic activity, changes in the rate at which the continents move across the surface of the globe (the rate of sea floor spreading), and changes in the magnetism of the Earth. The last discovery is particularly intriguing, not least because David Raup was one of several researchers who independently found this evidence of a roughly 30 Myr magnetic rhythm in the 1980s. It has to be said that this is far from being as clear cut as the periodicity found in the cycle of death itself, and the experts still argue about how reliable these analyses of the magnetic record are. But here is the basic idea.

From time to time, the Earth's magnetism fades away completely, and then builds up in the opposite sense, with the north and south magnetic poles replacing each other. This is purely a magnetic phenomenon – it does *not* mean, as some alarmists have inferred, that the Earth 'topples over' in space, turning upside down! The magnetic reversals are a very well-established fact of life on Earth, because magnetism at different epochs in the past is preserved in a kind of 'tape recorder in the rocks'. When some kinds of rock are formed in volcanic outbursts, with molten material flowing out from the volcanoes and setting solid, they become magnetised in line with the magnetic field that exists at the time. With successive layers of rock piled one on top of the other, geologists today can find out which way round the magnetic field of the Earth was at different times in the geological past by looking at layers of rock from different depths (different strata) – or, even more conveniently, by looking at strata which were laid down in layers in this way but have since been brought to the surface by geological upheavals, where they can be investigated relatively easily.

Paleontologists have been aware for some time that magnetic reversals are often accompanied by extinctions (large or small) of life on Earth. Several people have speculated why this may be so. One idea, for example, is that when the magnetic field is weak or non-existent, at the mid-point of such a reversal, charged particles from space (cosmic radiation) can penetrate to the surface of the Earth, instead of being deflected by the Earth's magnetic shield, and harm living things. But this is no more than a guess. Nobody really knows why life should be at risk during a magnetic reversal. Nobody knows, either, exactly how magnetic reversals happen. But since it is widely accepted that the Earth's magnetic field is generated by a flow of conductive material swirling around in the Earth's fluid core, it is not too difficult to imagine that some disruption in this flow could cause the magnetic field to be reversed. What better to cause such a disruption, Rampino suggests, than a violent strike on the Earth from the nucleus of a comet? That would surely set up swirling disturbances in the fluid core, easily sufficient to trigger a magnetic reversal.

Well, maybe. The argument is suggestive, rather than persuasive, and Rampino has no detailed explanation – no model – to account for how the phenomenon might work. The best evidence in support of the idea is that there is so much evidence that odd things do happen to the Earth every 30 Myr. There, Rampino is on more solid ground, making a persuasive case for the ubiquity of the 30 Myr cycle. And he is not alone. In 1993, for example, Ioannis Liritzis, of the Academy of Athens, published a major survey analysing the evidence for several claimed cycles of activity in terrestrial upheavals. He concluded[8] that the roughly 30 Myr cycle is of 'particular importance', showing

192

up not only in the mass extinction rhythms and the pattern of impact structures on Earth, but also in geochemical anomalies (including iridium layers, which have now been identified at five of the mass extinction boundaries in the geological record), volcanic activity, mountain building activity, and deposits of tektites. He also found the evidence for this cycle in magnetic reversals still persuasive, some ten years on from the original claims. The mean period Liritzis found for all this activity was 33 Myr, with an uncertainty of plus or minus 10 Myr, providing support for either Nemesis or the galactic carousel model, as you prefer. But the cratering record itself now seems to match the carousel model better than ever before. In the late 1980s, Eugene Shoemaker and his colleagues analysed the dates of 24 craters, reporting to the 1988 meeting of the American Association for the Advancement of Science that their arrival times showed a rhythm 30–32 Myr long, with at least one impact associated with four out of the five largest extinctions during the past 100 million years.[9]

It isn't difficult to link all of the geological changes to such cosmic impacts. Rampino points out that an object about the size of the nucleus of Halley's Comet (far smaller than Chiron), 'hurtling towards the Earth at thirty miles a second, would, on impact, release a million times more energy than the combined force of an average year's earthquakes'. Such a concussion could, he suggests, create a hot spot, either by the sheer force of the impact or by setting up convection currents in the fluid material beneath the Earth's crust. This would lead to a chain of volcanic activity, mountain building, movement of the crustal plates, and changes in both sea level and climate. 'Somewhat as a rear end collision abruptly accelerates a slow-moving automobile, a great impact would jog the progress of geological change,' he says.

Together with evidence that the most recent episode of mass extinctions and of impact cratering occurred between one and five million years ago, this makes an extremely attractive package; it links many processes which have shaped the face of the Earth and influenced the evolution of life on Earth to our place in the Universe at large. Just as it would be a remarkable coincidence if the carousel effect just happened to have the same rhythm as the extinction record, so it is even harder to see how all these 30 Myr periodicities could have arisen by different mechanisms and be entirely independent of each other. Leaving aside the still-contentious case of the magnetic reversal cycle, Rampino and Stothers calculate that the probability of the geological, tectonic and cratering periods being accidentally equal to each other is less than one in 10,000 – and that still doesn't include an allowance for the probability that the same cycle also turns up in the fossil record. Further, just as a flaw in the Nemesis model is that the orbit of the Death Star would not be stable enough to match the rhythm found in the fossil data, so the stability of the galactic carousel effect, repeating steadily with the same rhythm for many hundreds of millions of years, is a big point in this theory's favour. It is all of that which makes the galactic carousel our best buy to account for the 30 Myr cycle of death.

Whichever way you look at it, though, this cycle is not something that we should worry about as far as the immediate future of life on Earth is concerned. If the Nemesis idea is correct, we probably have another 13 Myr to wait before a rain of comets falls upon the inner Solar

System. If Rampino is right, we are safe for another 30 Myr or so. But there is still cause for concern about the possibility of lesser impacts, on the scale of the Tunguska Event. Nobody suggests that comets *only* fall in to the inner part of the Solar System at intervals of 30 Myr or so; just that many more comets come our way then. A supercomet like Chiron could come our way at any time, breaking up like Shoemaker–Levy and scattering its fragments like buckshot across the path of the Earth through space.

And history suggests that we are, indeed, living in the aftermath of the break-up of such a supercomet (possibly even a straggler from the shower shaken loose from the Öpik–Oort Cloud in our latest passage through the plane of the Galaxy). We mentioned some of the astronomical evidence for this in chapter five. Now, the time has come to look at the consequences of that break-up for the recent – very recent – history of life on Earth.

Notes

1 *Volume 35*, p. 1 (1994).
2 *The Nemesis Affair*, p. 114.
3 *The Mass-Extinction Debates*, p. 143.
4 This would have come as no surprise if Raup and Sepkoski had known more about astronomers. In astrophysics, theorists are desperate for new bits of data on which to build their theories, and are used to coming up with several crazy ideas for each new observation that is made.
5 Told by Muller in his book *Nemesis: The Death Star*.
6 Rampino explained the Rampino/Stothers model of the cycles of extinction particularly clearly in an article in *The Sciences*, July/ August 1987, published by the New York Academy of Sciences, from which quotes by Rampino in this chapter are taken.
7 Volume 255, p. 607.

8 *Quarterly Journal of the Royal Astronomical Society*, volume 34, p. 251.

9 We can't resist pointing out that if the period of the cycle of death really is 30 myr, it could still be explained by a Death Star in a slightly different orbit from the one proposed by Muller and his colleagus; but since Nemesis probably does not exist, this is something of a red herring.

IX

FIRE ON EARTH

It has taken historians a long time to appreciate the importance of relatively minor fluctuations in climate on the activities of humankind and the development of civilization. Today, although the study of the impact of climatic change on history is at least respectable (although many historians still balk at the idea of such investigations), all too often even these studies are restricted to the broad sweep of climate, the kind of changes that take us into, and out of, full Ice Age conditions. It is still rare to find anyone prepared to consider the possibility of what might be called a micro-influence of climatic change on history – the influence of the kind of short, sharp shocks to the climate system that would be produced when a pulse of Tunguska-type events occurred within a few decades or centuries.

The limited studies that have been made suggest that such pulses of relatively minor cometary impacts with the Earth, and the associated spread of dust in the atmosphere of the Earth, may indeed have caused several climatic blips during recorded history, influencing the course of such not-insignificant events as the fall of the Roman Empire.

We looked at the broad influence of climate on history in an earlier book, *Being Human*. We don't want to go into all the details again here. But before we look at the way in which impacts from space may have brought first fire and then ice to the Earth in recent millennia it seems a good idea to set the scene by painting in the broad outline of the link between climate and history. It is, of course, no easy task to reconstruct the patterns of past climate. Sources of evidence include the changes in the width of tree rings in old wood, changes in the kind of pollen deposited in the sediments of lake beds, old written records, geological evidence of the advance and retreat of glaciers, and so on. If you want to know more about these techniques, and about the reconstructions of past climates that they make possible, Hubert Lamb's book *Climate, History and the Modern World* is a good place to start. The broad picture that emerges is unambiguous, although there is still debate about some of the details.

Between 15,000 and 10,000 years ago, the world warmed out of the most recent Ice Age, which we discussed in chapter seven. As the ice retreated, the meltwater from the glaciers caused sea level to rise by 50 metres in that 5,000-year interval, and by a further 40 metres between 10,000 and 5,000 years ago. As always throughout history (and prehistory), many of our ancestors lived on the shoreline, in particular around the Mediterranean, and these huge rises in sea level would have had a dramatic effect on their lives. With water rising at half a metre each century, and the world getting warmer, people were forced to adapt and move to take advantage of the changing situation almost literally from generation to generation. It is possible that although the climate improved (by most human criteria) at the end of the Ice Age, the human population actually declined at first, as people were displaced by the rising waters and had to seek new homes and new lifestyles. It is also more than possible that the worldwide legends of a great flood date from those days. The world was at its warmest and wettest in the interval from 7,000 to 5,000 years ago, encouraging a human population explosion, once the initial problems of the rising seas had been overcome.

It is no coincidence that during this climatic ameliora-tion agriculture was developed independently at three locations around the world – in the fertile crescent which includes the valleys of the Tigris and Euphrates rivers in the Middle East, in China, and in Central America. The settled lifestyle of the farming communities, and the increased amounts of food that farming made available, encouraged the growth of the first cities. About 5,000 years ago, however, there was a major, but relatively short-lived, climatic setback. Temperatures fell dramatically, and stayed low for about 500 years before recovering, equally quickly, to about their previous values. It was as if the steady sweep of gradual climatic change had been interrupted by some outside influence.

Then, the broad sweep of climatic change caused the world to cool slightly, in a more leisurely fashion. By 3,000 years ago, the glaciers had advanced again, further than they have ever been since. Hubert Lamb, of the University of East Anglia, has suggested that the memory of these icy centuries in northern Europe may have formed the basis of the Norse legend of the end of the world – Ragnarok, the twilight of the gods. The Ragnarok story tells of the coming of the Fimbulvinter, a run of three severe winters in succession with no summer in between, associated with a great fire. Lamb has suggested that the great fire, like the severe weather, was a real event and that it occurred when forests that were killed and dried out during the long winter burned; but, as we shall soon see, there may be another explanation for this link between ice and fire.

As the world froze in the north, it dried out nearer the equator, with deserts expanding as a result of the same climatic shift. The only place left for European civilization to develop[1] was in the middle, around the Mediterranean. Climatic changes played a direct part in ensuring that Egypt, Greece and Rome were the great powers of early times, which carried the growth of civilization forward. It was Rome that spread civilization northward and westward across Europe not just because the Romans were more conveniently located geographically to do this than Egyptians or Greeks, but also because the severe cold spell ended around 2,500 years ago (around 500 BC). The growth in importance of Rome from a single city-state, and the spread of the Roman Empire across Europe, began around that time, and were considerably assisted by the amelioration of the climate. The Romans moved north as the weather warmed, taking their vines with them and planting them successfully in places where they would never have thrived 500 years earlier.

But it lasted for less than a thousand years. There have been many suggestions why the Roman Empire in Europe fell (the eastern Empire, Byzantium, survived much longer), but one important factor, which we discuss in more detail later, must have been the spell of severe climatic conditions that occurred in and around the fifth century AD. There were certainly other reasons for the decline of the Empire, including corruption and bad management. But the eastern Empire was also often corrupt and badly managed, and still survived for almost another millennium. A major factor in the fall of Rome was that it was pushed – overrun by hordes from the north. And what better reason would those hordes have to move south than difficulties at home caused by a deteriorating climate?

In the second half of the first millennium AD, the climate improved again in Europe, and Norse voyagers travelled the then-known world, even reaching America. But their activities were curtailed as the ice returned to the north once again, blocking off their sea routes to the west. In the 14th century, Europe was hit by a spate of disasters – crop failures, starvation, and the plague known as the Black Death. It is now clear that this is indeed the order in which the disasters struck, and that preceding all of them there was a sharp downturn in climate. It was climatic change that caused the crop failures, starvation that weakened the population, and then the Black Death that wreaked havoc on the weakened, starving people, leading to whole villages being depopulated and abandoned, right across Europe. The disturbed populations that survived were involved in all kinds of civil unrest, including witch-hunts and uprisings in both England and France.

After a slight recovery from these extreme conditions, cold weather returned in full force in the 17th century, which has become known as the time of the 'Little Ice Age'. Plague also returned, and so did civil unrest – in England reaching the extreme form of all-out Civil War. People were cold, hungry and desperate for change when the Little Ice Age was at its worst, and that contributed in no small measure to the success of Oliver Cromwell's revolution. It was also the reason that Scotland agreed to the political union with England in 1707. In the last decade of the seventeenth century, roughly half the population of Scotland died as a result of famine and disease, and the country was desperate for the food aid that the union brought.

It was a similar story across Europe, and even in North America, as we explained in *Being Human*. But this is enough background to be going on with, and to make the point that is relevant here. There have indeed been episodes of very cold weather affecting human populations at intervals of a few hundred years over the past few millennia. We discussed some of the standard explanations for these climate shifts – in particular, the reasons for the broad sweep of climatic change over the millennia – in our earlier book. But there is now a growing weight of evidence that one influence we neglected to mention then, impacts from space, played a part in at least some of these turning points in history. The leading proponents of this idea today are Victor Clube and Bill Napier, whom we have already met.

The obvious question is why, if climate-induced disaster (for which there is plenty of evidence) was caused by cometary impacts in the past, the ancients didn't leave us records describing those cometary impacts. Clube and Napier offer persuasive arguments that those records do indeed exist, but have not always been recognised as descriptions of real events affecting the Earth. They are the myths and legends of warring gods, that carry a similar message from many unrelated cultures spread right around the world.

These ideas were developed by Clube and Napier in their books *The Cosmic Serpent* and *The Cosmic Winter*, both of which we recommend. The basis for their re-interpretation of the old stories is the evidence, which we discussed earlier in the present book, that the amount of cosmic debris in the inner part of the Solar System today (in particular, the material associated with the Encke stream and the Taurid meteorites) is so great that impacts on the scale of the Tunguska Event and upwards are far from being rare in recent Earth history, and in some centuries must have been quite common. Within the past 5,000 years, since the end of the great rise in sea level associated with the end of the latest Ice Age, there are estimated to have been about 50 impacts, each with an energy between 1 and 100 megatonnes (for comparison, the Tunguska Event lies at the top of this range); five impacts each with an energy in the range from 100 to 1,000 megatonnes; and a 50–50 chance that there has been one impact with an energy in the range from 1,000 to 10,000 megatonnes. But these estimates represent a 'steady state' calculation of the frequency of the kind of events that would be scattered at random throughout the centuries. When the Earth encounters a comet in the act of breaking up, or crosses the stream of debris from such an event, a burst of much more severe activity can be expected.

The break-up of the kind of supercomet we discussed earlier can hardly have gone unnoticed by anyone alive on Earth at the time. We know for certain, from ancient Chinese records, that Halley's Comet was seen in 1058 BC as a much brighter object than Venus in the night sky. Astronomers measure brightnesses in terms of magnitudes, a curious system of units in which, for historical reasons, fainter objects have larger positive numbers, and very bright objects have negative numbers. Each step in the magnitude scale corresponds to an increase in brightness of 2.5 times. So when astronomers tell you that Venus has a brightest magnitude of –4.4, as seen from

Earth; and that in 1058 BC Halley's Comet (as far as can be judged from the description that has come down to us) had a brightest magnitude of –7.7, as seen from Earth; this means that, since the difference between –7.7 and –4.4 is roughly 3, the comet was 2.5^3 times brighter than Venus, and 2.5^3 is 15.6. The comet was actually *fifteen times* brighter than Venus at its brightest!

Although this was the earliest recorded observation of Halley's Comet, Clube and Napier estimate that, given the amount it has faded since 1058 BC, around 2000 BC it must have been about a hundred times brighter than Venus. The original supercomet which broke apart to produce (among other things) Encke's Comet and the Taurids would have been about six times brighter still – about as bright as the full Moon and amply bright enough to cast shadows at night.

There is, as we have said, no way that people could have failed to notice such an object, looking like:

> An intense yellow spot of light surrounded by a circular coma probably larger than the full Moon, with a tail stretching across a large part of the sky . . . graduating from bluish white near the nucleus to a deep red in colour, the whole being finely structured.[2]

Where are the descriptions of such objects in ancient literature? The first place Clube and Napier point us towards is the Bible, where we read in Genesis (15:17) that 'When the sun went down, and it was dark, behold a smoking furnace, and a burning lamp'. While in I Chronicles (21:16) the Angel of the Lord is described as appearing to David as 'standing between the earth and the heaven, having a drawn sword in his hand stretched out over

Jerusalem'. They interpret this as describing a spectacular comet, with the face of the Angel represented by the coma while the curving tail represents the sword.

A circumstantial piece of evidence in support of the influence of cometary impacts in ancient times is that all cultures seem to perceive comets as harbingers of doom – in AD 66, for example, Halley's Comet was seen as a warning to the Jews that Jerusalem would be destroyed. Why would people who were every bit as intelligent as us, even though less advanced technologically, imagine that comets brought death and destruction, if that death and destruction had not been seen, if not with their own eyes then with the eyes of their own recent ancestors?

It is also curious, to say the least, that the Latin word for stars is *sidera*, while the Greek word for iron is *sideros*. The two words have a common root, pointing to an association in ancient times between stars and lumps of iron. The only lumps of iron the Ancient Greeks were likely to be aware of were meteorites, which they could have seen falling to Earth as shooting stars. More evidence that people in ancient times were aware of the links between the Earth and the heavens – better aware of these links, in fact, than more modern scientists, who dismissed the idea of rocks falling from the sky as utter nonsense as recently as the 18th century.

The few identifiable descriptions of comets in ancient texts often refer to them as 'hairy stars', making an analogy with the long hair of a woman, streaming behind her in the wind. This encourages researchers such as Clube and Napier to make connections between goddesses named in ancient stories and comets – especially when there is other evidence to support the identification. In Babylonian mythology, for example, there are three great gods,

Shamash, Sin and Ishtar. The first two are known to be identified with the Sun and the Moon, respectively, but the identification usually made for Ishtar is with the planet Venus, which is, as far as its appearance on the night sky is concerned, literally a pale imitation of the other two. Yet Ishtar ranked equally with the other two main gods in the Babylonian pantheon. A comet as bright as the Moon would be a much better candidate, and as the comet faded away the name itself might easily have been transferred to the planet Venus.

Direct evidence for the influence of fragments of comets on the Earth comes from ancient descriptions of thunderbolts, which were clearly not just the crashing sound associated with lightning. The Roman, Lucretius, for example, refers to how 'with the thunderbolt's heavy freight of fire and wind, it trails in its wake a murky tempest big with levin-bolt and blast'[3]: an evocatively familiar image to anyone who has read the descriptions of the fall of the Tunguska meteorite. But in the decades that were recent history to Lucretius (who lived in the first century BC) such events must have been far more common than they are today. In trying to interpret the mystery of why such thunderbolts seemed to strike at random over the surface of the Earth, he cast doubt on the conventional wisdom that it was the work of the gods, and in particular of Jupiter, pointing out that it made no sense for the gods to hurtle their thunderbolts into the desert or the open sea, rather than down upon the heads of sinners.

Yet, the random nature of these acts of terrorism can be seen as an argument in favour of the idea that the gods were comets. One of the key features of Greek and Roman mythology is the capriciousness of the gods, often acting on whim, rather than for any logical reason. Using the account of the story of the origin of the Universe and the history of the gods given by the Greek poet Hesiod (who lived in the eighth century BC) in his *Theogony*, Clube and Napier offer a deliciously persuasive interpretation of the way religion itself developed and changed as events in the heavens unfolded. Olympus, the home of the gods, is seen as a great comet, moving across the sky, where Zeus reigns, hurtling his thunderbolts earthward from time to time. Later, a huge battle takes place in the heavens between Zeus and the Titans, identified as a shower of new comets (perhaps produced in a break-up like that of Shoemaker–Levy in 1994). Then, Zeus:

> came from heaven and Olympus. Thick and fast, the thunderbolts, with thunder and lightning, flew from his stout hand and they made a holy flame roll along, as they came in quick succession. The life-giving earth blazed and crashed all around, and all around immense woods crackled loudly in the fire . . . the gleaming brilliance of the thunderbolt and lightning blinded their eyes . . . the winds stirred up earthquakes, dust, thunder, lightning and smoky thunderbolts, the arrows of the great Zeus, and carried shouts and war-cries . . .[4]

All this is conventionally interpreted as a work of fiction; but it reads like an accurate description of a series of Tunguska-like events. Even if it were fiction, what put the idea into the writer's head? All fiction, after all, is based on fact! The story goes on to tell us that, after all this excitement, one god/comet was left dominating the night sky, until it too faded away and became invisible.

The evolution of religious ideas often follows such a pattern, not just in the Graeco-Roman world but elsewhere: a family of gods war amongst themselves and are replaced (or succeeded) by a single god, who is at first closely involved with human activity in a highly visible, 'hands-on' way, but later becomes an invisible, 'behind the scenes' manipulator of human destiny. Five thousand years ago, gods were described as real entities in the sky, who fought and squabbled amongst themselves. A thousand years later, the image was of two great gods in heaven, one good and the other evil, with their conflict leading to fall-out on Earth in the form of thunderbolts. Within another thousand years, only the benevolent god remained, and even he then faded from view. Could it be, as Clube and Napier suggest, that all of this is based on what our ancestors could see going on in the sky above them?

Where else do we see similar evidence of cometary activity? We have already mentioned the Norse mythology. Change the names of the gods involved, and Hesiod's description, quoted above, would apply just as well to the Ragnarok story. Both Norse and Greek writers seem to have been describing the same event, seen as trouble among the gods in the sky bringing trouble down upon the surface of the Earth. And if they are separate descriptions of the same event, that makes it even harder to see how the stories could not have been based on fact. Another aspect of the Norse mythology is the appearance of the World Tree, Yggdrasil, which was supposed to lie between heaven and Earth, and which could also fit the appearance of a comet. And the association between fire and ice in the story of Ragnarok is tellingly like the image of cosmic winter that has become a respectable scientific scenario of the events associated with the death of the dinosaurs.

There is no need to labour the point. It is certainly worth considering the possibility that events described in ancient writings (including, for instance, the destruction of Sodom and Gomorrah) were indeed eye-witness accounts of what they purport to describe. They might not be stories made up out of whole cloth just for the fun of it, even if the interpretation the ancient writers put on those observations may look strange to modern eyes. The next step is to see how those descriptions fit into the pattern of history, and also into the pattern of the Earth's known encounters with the Encke stream.

History really began at the time of the climatic decline around 5,000 years ago, following the warmest post-Ice Age millennia. This was the time when some societies became rich enough, and well organised enough to undertake huge projects such as building the Egyptian pyramids and great irrigation schemes; it was also the time when writing was invented, largely in response to the bureaucratic need to keep track of what was going on in these projects. Such activity was made possible by agriculture, which meant that although some people had to toil in the fields, the food they produced was sufficient not just to feed themselves and their own families but to provide a surplus for workers who were engaged on projects which did not immediately put food into their own mouths.

It is possible that the climatic decline actually triggered this upsurge in the development of civilization. A plausible scenario would be that after centuries of ease in a friendly climate, the change for the worse forced people to develop irrigation systems and better organised farming, with a centralised distribution network. When the climate eased again the existence of this infrastructure made other great projects, such as pyramid building, possible. This can only be speculation, but in any case it is tangential to our story. What is definitely not speculation, and is much more relevant here, is that it was at about the time of this climatic decline that pharaohs who ruled in different parts of Egypt began to identify themselves with gods, and people began to see gods as playing a major part in everyday life. Like other creation myths, the Egyptian story describes a great battle between the gods which ends in the triumph of a single powerful god. The story can easily be interpreted as describing exactly the same pattern of cometary activity that Clube and Napier argue lies behind the Greek creation myth. And, once again, there are stories of cataracts of fire and occasions when thunderbolts or blasts from the gods lay waste to the land.

Clube and Napier link these stories of fire from the sky with fire festivals that are still held in many parts of the world.[5] Given the importance of fire to early humans, it isn't too surprising that these festivals are common around the world. But it is rather odd that these festivals often tend to be held either early in November or in mid-June. The summer festival is fair enough – it could be explained as just that, a celebration of the 'longest day'. But why early November, and not mid-winter, a time when people might be expected to want the comfort of large bonfires? Of course, there are mid-winter festivals as well, but the early November fire festivals are important enough to have been taken over for other purposes down the centuries, leading, for example, to Halloween and the English bonfire night of 5 November. The Chinese Feast of the Dead, which involves bonfires and lighted candles, and the Japanese Feast of Lanterns also take place at this time of year. Could it really just be a coincidence that these fire festivals are held at the times when the Earth crosses the Encke stream, and meteor (and meteorite) activity is at a peak? The meteors may be relatively unspectacular now, but at certain times in the past, including 5,000 years ago when the climate blipped and civilization started on the road to centralised organisation, they would have been much more impressive, even threatening.

We have already mentioned that gradual climatic change may have assisted the rise of the Roman Empire. Clube and Napier suggest that the Empire's decline was hastened, ushering in the Dark Ages, by a much more dramatic, short-lived climatic event, linked, once again, with disastrous thunderbolts striking the Earth. They point in particular[6] to a chronicle written by a British priest, Gildas, less than 200 years after the event.[7] The picture that emerges is of a Britain that was still a well ordered, civilized state (in the Roman sense, but no longer ruled by Rome itself) as the middle of the fifth century AD approached, but which was then overwhelmed by what Gildas calls 'the ruin of Britain' in the form of:

> the fire of righteous vengeance [which] wasted town and country . . . Horrible it was to see the foundation stones of towers and high walls thrown down bottom upwards in the squares, mixing with holy altars and fragments of human bodies.

At the same time, there is evidence of a dramatic deforestation of Britain, which could have been caused by huge forest fires (indeed, it is hard to envisage any other cause). The population was indeed devastated, and so many survivors moved south into the nearby part of France that to this day the region is known as Brittany.

211

The scale of the disaster that struck Britain in the fifth decade of the fifth century AD was far worse than anything which happened elsewhere in Europe around that time, and the region seems to have been particularly unlucky in suffering a major impact from space. But there may well have been similar events during the fifth and sixth centuries in other parts of the world, where there were no chroniclers on hand to record what had happened. Such disasters striking in Siberia, for example, or even northern Europe, could have gone unrecorded historically, but still played a part in causing the population movements which put pressure on what was left of the western Roman Empire. While even impacts that occurred in, say, North America would also have played their part in affecting the global climate.

There was indeed a general deterioration of climate worldwide at around this time, which could neatly be explained by dust, associated with the incoming comet fragment(s), spreading through the atmosphere and acting as a sunshield. Dust layers are often found in ice cores drilled from the glaciers of Greenland and Antarctica, usually in layers of ice that can be dated as having been laid down during cold spells on Earth, and this is no exception. The conventional interpretation of these dust layers is that they are a result of volcanic activity, but comet dust, as we shall shortly discuss, would do just as good a job of global cooling, and would leave similar traces in the ice.

The icing on the cake of the suggestion that Britain was hit by cosmic impact(s) in the fifth century is that observers in both Spain and China reported seeing an unusual comet in AD 441. The comet itself clearly did not strike the Earth, or the devastation would have been more widespread (and we probably wouldn't be here today to puzzle over it). But just a piece of comet, or a few Tunguska-sized pieces, broken off and falling on Britain, would be enough to account for the events described by Gildas. And, as you may already have guessed, the Earth passed right through the dust trail associated with Encke's Comet and the Taurid meteors at around this time.

It was only after these events had taken place, causing the collapse of the Roman-style civilization in Britain and depopulating the land, that the Anglo-Saxons moved in – not fighting their way ashore and destroying the old way of life, but simply taking advantage of the collapse that had already occurred. But why did these invaders from the continent of Europe want to move in to the island of Britain? It is at least worth considering the possibility that they were being pushed to seek new homes by the pressure of population movements in the heart of the continent, which may themselves have been triggered by cosmic impacts striking far outside the borders of the Romanised 'civilized' world of the time. Barbarian population movements certainly contributed to the fall of the Roman Empire, and Clube and his colleague David Asher, who now works at the Anglo-Australian Observatory, argue that the barbarian movements were caused by impacts from space.

The Dark Ages that followed the collapse of the Roman Empire provide the most extreme example of the kind of terrestrial catastrophe that may have been triggered, or at least influenced, by cosmic impacts since the time of Christ. But it is now clear that there have been many minor climatic fluctuations in the past few thousand years, and, naturally, we know more about the more recent ones.

This is why the Little Ice Age that brought such severe weather to Europe in the 17th century is often referred to in those terms – as 'the' Little Ice Age. In recent decades, however, climatologists and historians have discovered a repeating pattern of little ice ages occurring at intervals of roughly a thousand years, right back through the past 10,000 years (the interval, known as the Holocene, since the end of the latest Ice Age). The evidence has been summarised by Jean Grove, of the University of Cambridge, in a *tour de force* book called *The Little Ice Age*, published in 1988.

The rhythm that Grove has drawn attention to is not a regular pulsebeat producing bursts of cold decades at precisely spaced intervals throughout the Holocene. Rather, it is a tendency for cold conditions to recur after longer intervals of relative warmth. This tendency shows up more or less strongly at different times in the Holocene, not least because ten thousand years is such a long interval of time that climatologists have to take account in their calculations of the way changes in the movement of the Earth in its orbit (the Milankovitch processes) affect the balance of the seasons. At the beginning of the Holocene, there was a greater contrast between the Northern Hemisphere seasons, with more heat from the Sun than today arriving at the Earth in June, July and August, and less in November, December and January.[8] As we mentioned in chapter seven, the increased summer warmth must have helped to melt back the great ice sheets that had covered the northern continents during the preceding Ice Age.

This also meant that at that time, in the early Holocene, the southern hemisphere experienced relatively cool summers (and relatively mild winters, of course, so that the average amount of heat received from the Sun over the entire year stayed the same). It is cool summers that allow glaciers to grow, because it is always cold enough on high mountains to make more ice in winter, and what matters is how much (or how little) of the winter's new production of ice melts back in summer. So, early in the Holocene, little ice ages were more pronounced in the southern hemisphere, while in the past couple of millennia they have been more pronounced in the north.

There is no doubt, though, that once these factors have been taken into consideration the little ice ages are seen affecting the whole world. The evidence for the cold spells, in the form of advancing glaciers, has been found in Scandinavia and the Caucasus mountains of Europe, the Himalayas, both North America and South America, and New Zealand.

Several possible explanations for the phenomenon have been put forward. One suggestion is that it may simply represent the random fluctuations of the climate system (another example of chaos at work). Other people link (or attempt to link) the climatic changes with changes in the Earth's magnetism, or with changes in the activity of the Sun. But one of the best possibilities – what Grove calls 'an attractive possible explanation' – is that the amount of dust in the atmosphere increases from time to time.

When Grove called this an attractive explanation, she was specifically referring to dust (and other material) put into the atmosphere by volcanic eruptions. There is no doubt that upsurges in volcanic activity do coincide with at least some of the little ice ages of the past few thousand years, and there is equally no doubt that major volcanic eruptions can influence the climate – these ideas were dramatically put to the test when the volcanic Mount Pinatubo, in the Philippines, erupted on 12 June 1991.

That single volcanic eruption threw about 20 million tonnes of sulphur dioxide into the stratosphere, along with comparably large amounts of very fine dust (rather like talcum powder). The sulphur dioxide reacted with atmospheric water to make tiny droplets of sulphuric acid in the stratosphere (a kind of stratospheric smog). Together, the acid droplets and the dust screened out about two per cent of the incoming heat from the Sun, preventing it from reaching the surface of the Earth. The average temperature over the whole world fell by between a quarter and one half of a degree Celsius, and did not recover until late in 1994, when the polluting haze had cleared from the stratosphere (the stratospheric smog from Pinatubo also, incidentally, exacerbated the problem of ozone depletion over both Antarctica and the Arctic). Dust in the atmosphere undoubtedly does cause global cooling, and volcanoes do blast dust into the atmosphere.

But as Grove also points out, 'there is not a one-to-one relationship between volcanic eruptions and frost events'. Sometimes, there are eruptions (like Mount St Helena, in Washington State, in 1980) which throw dust out sideways instead of upwards, devastating the local environment but not producing a stratospheric global dust veil. Sometimes, there are severe cold spells that do not seem to be associated with volcanic activity. So, clearly, volcanism is not the whole story. There must be something else disrupting the climate system from time to time as well, something which produces a very similar pattern of climatic changes to those produced by volcanic dust veils.

This leaves the door open for the idea that some of these little ice ages may have been produced by the same kind of mechanism, dust in the atmosphere, but originating from a different cause, the impacts of cometary fragments with the Earth. In their enthusiasm, proponents of this idea, including Clube and Napier, sometimes try to argue that all of the little ice ages have been produced in this way. We think that this is going too far. The link between volcanic activity and global cooling is so clear-cut that there is no doubt that some little ice ages are primarily caused by upsurges in volcanism. And one of us, in particular, has a weakness for the idea that changes in solar activity may also play a part. But what really matters about the study of the links between volcanic activity and climate is that they show just how sensitive the weather machine is to the spread of fine dust in the stratosphere.

The weather machine is not some great juggernaut which proceeds inevitably in a certain direction, like a huge locomotive hurtling down a railroad track. Its direction can be switched quite dramatically (in terms of the resulting impact on human activities) by relatively small outside influences. There is no doubt that comet fragments do strike the Earth from time to time (remember the Tunguska Event!). And such impacts can spread dust into the stratosphere both from the explosion when the object hits the ground and from the trail it leaves as it thunders through the air. So, it is inevitable that a cluster of such impacts occurring in the space of a few decades will produce the same kind of global cooling as an outburst of volcanic activity.

It is even possible that 'the' Little Ice Age was produced in this way. In a lecture given at Green College, Oxford, early in 1995, Clube drew attention to a proliferation of pictures in England illustrating people looking at heavenly fireballs from about 1640 to 1680, and linked this with an upsurge of concern about the end of the world – what he calls 'millennial' fears – at that time. He links all of this to the kind of breakdown in society typified by the English Civil War, and also points out that Chinese records show an upsurge in the occurrence of 'fireballs' at around that time. Clube's interests now focus on the public response to such clusters of meteors and meteorites – not just their response to the physical problems caused by the impacts, but the psychological response to these 'heavenly' portents. This sociological/psychological connection lies outside the scope of the present book. But whatever the connection between the appearance of lights in the sky and the spread of civil disorder on Earth, it does seem that there really were more lights in the sky than usual during the 17th century.

The evidence of a long-term link between dust in the atmosphere and global cooling comes, as we mentioned earlier, from traces of the dust left in the layers of ice in the Greenland and Arctic ice sheets. It was natural that when these dust traces were first discovered they should be seen as supporting the volcanic view of climate change. In fact, at least some of these dust traces may represent cosmic winters brought about by impacts from space. The intriguing question is, which of the other little ice ages, and the associated upheavals caused for humankind, can be linked with impacts from space?

In our view, the strongest cases made by Clube and Napier (and by Clube and David Asher) are for impact events associated with the Dark Ages (occurring roughly 1,500–1,600 years ago) and the climatic blip around 5,000 years ago (which they link with an upsurge in the growth of civilization and the origins of worldwide mythologies about gods fighting in heaven). And one reason why these look like such good candidates is that the researchers have an explanation for why the events should have occurred at those particular times.

As we saw in chapter five, backtracking from the present orbits of objects in the Taurid stream, including Encke's Comet, suggests that a major break-up of the parent body occurred about 5,000 to 4,500 years ago. The debris from such a break-up does spread slowly along the entire orbit of the stream of material, but it stays concentrated in a certain part of the stream (corresponding to the original orbit of the object that broke up) for a long time. So although the Earth may pass through the extended dust trail every year (producing, for example, the Taurid meteor showers themselves), the greatest risk of multiple Tunguska-type impacts occurs when it is passing through the densest part of the stream. This doesn't mean that the greatest risk occurs in just a single year; rather, because the material is spreading out along the orbit, there will be an increased risk of impacts on the scale of the Tunguska Event for several decades, followed by a period of relative calm lasting for centuries or millennia before the orbits of the Earth and the densest part of the stream coincide in just the right way again.

These 'nodal intersections' recur at intervals of roughly 2,500 to 3,000 years – although the interval between one danger period and the next does not stay exactly the same, because the orbit of the stream of cometary debris is changed slightly over the millennia by the gravitational influence of the planets, especially Jupiter. Both of the two best examples of changes on Earth brought about by this kind of impact from space discussed by Clube and his colleagues fit this pattern – and it doesn't take too much calculation to appreciate that, if they are correct, the next period of increased risk from cosmic impacts will be in about a thousand years' time, around the year 3000. If the calculations made by Clube and his colleagues are correct, at that time the increase in the occurrence of cosmic impacts will be sufficient to produce, on average, one impact as big as the Tunguska event in each area of the world the size of England in a span of about a hundred years.

This doesn't mean that the impacts will actually be spread out evenly across the globe. As seems to have happened in the fifth century, some regions will suffer multiple impacts, while others will get off scot free. But even today (let alone in the year 3000), there are very few regions, on land at least, where even a single such impact would cause as little direct harm to human populations as the Tunguska Event did in 1908. And nor, as the example of the Tunguska Event itself highlights, does the fact that the greatest risk will be in a thousand years from now mean that there will be no risk before then. There are undoubtedly stray fragments of cometary debris scattered here and there, like islands, in the stream, and one of those stray fragments could be in an orbit that will intersect the orbit of the Earth any year now. Is there anything we can do about this? Is there anything we *should* do about this?

Notes

1 We concentrate on European civilization here both because it is our own heritage and because it is the one for which historical records are most easily available.
2 *Cosmic Serpent*, p. 154.
3 See the translation of Lucretius' *The Nature of the Universe* by R. E. Latham (Penguin, 1951).
4 Quoted in *The Cosmic Serpent*, p. 184.

5 *The Cosmic Winter*, Chapter One.

6 *The Cosmic Winter*, Chapter Seven.

7 See M. Winterbottom, *Gildas* (Phillimore, Chichester, 1978), as quoted in *The Cosmic Winter*.

8 We should stress that what changes is the amount of heat *arriving at the Earth* in the appropriate months, not the amount of heat *leaving the Sun*. The change happens because the Earth is a little closer to the Sun at one end of its orbit than at the other; 10,000 years ago it was at perihelion in northern summer, today it is at perihelion in northern winter.

X

WHAT TO DO

On 15 March 1994, an asteroid 10 metres in diameter passed by the Earth at less than half the distance between the Earth and the Moon. The object, dubbed ES1, was detected by one of the few telescopes dedicated to the search for asteroids that come near the Earth: the 0.9 metre Spacewatch instrument on Kitt Peak, in Arizona.[1] It was first detected only the previous day, when it was just 2.2 million km outside the Earth's orbit, and its orbit was determined by Brian Marsden, of the Harvard Smithsonian Center for Astrophysics. During the 1990s, about one such close encounter has been observed each year. But, given how little telescope time is devoted to looking for such objects, the experts believe that this amounts to no more than one in a hundred of the objects like ES1 that are actually passing close by the Earth. Such 'near hits' probably happen once or twice a week, but pass unnoticed because there is no large-scale survey of the skies going on to seek them out.

Even if we could not do anything to divert an incoming meteoroid from its course, it surely would be wise to set up such a monitoring system, so that everyone could know the origin of explosions on Earth or in the atmosphere of our planet. A couple of months before ES1 passed us by, on 1 February 1994, a fireball 20 km above the Pacific Ocean was detected by six US spy satellites. Later analysis of the data showed that the fireball was produced by the disintegration of an incoming object with a mass of more than a thousand tonnes, producing an explosion equivalent to 100 kilotonnes of TNT. But at the time, the first reaction of the military was that it might be a nuclear explosion, and the US President, Bill Clinton, was woken by the defence staff to discuss the implications.

In this case, we were lucky – the explosion happened over an unpopulated part of the globe, at a time of little tension between the military superpowers. But remember that something similar happened over western North America on 1 October 1990. At that time, the tense situation in the Middle East that led to the Gulf War was building up. If that explosion had occurred over Iraq, say, it might well have been interpreted as an attack from Israel, prompting a military 'retaliation'.

A couple of years later, in the northern hemisphere autumn of 1992, there was a flurry of attention in the media for another cosmic disaster story, presented (with extreme amounts of hype) as a real likelihood that a comet known as Swift-Tuttle would strike the Earth in the year 2126. Although no astronomer ever made such a prognostication, the story of Swift-Tuttle does highlight both some of the dangers and some of the difficulties faced by anyone trying to set up an early warning system.

Swift-Tuttle gets its name from the two astronomers, Lewis Swift and Horace Tuttle, who independently discovered the comet in 1862. It is the parent body of the Perseid meteor stream, and since the Earth passes through the material that produces the Perseid shower (which means we see this major meteor shower every August), it has long been clear that the comet itself could come close to the Earth from time to time. But until a few years ago, nobody knew the exact orbit of Swift–Tuttle.

The observations from 1862 suggested that the period of the orbit was about 120 years, but it turns out that they were not very accurate. Back in 1973, in a paper published in the *Astronomical Journal*,[2] Brian Marsden suggested that a search for Swift–Tuttle might be made in the early 1980s. But in that paper he also pointed out that it was curious that there were no other known records of such a bright comet. In 1862, Swift–Tuttle was unmissably bright, and should have been at least as bright (as seen from Earth) on one orbit out of five. Indeed, Marsden pointed out 'near collision with the Earth would take place if the comet were at perihelion in July'. So he tentatively suggested that Swift–Tuttle might indeed have been seen before, and in particular that it might be identified with a comet seen in 1737. That would mean that the orbital estimates were wrong, and that the next visit of Swift–Tuttle to the inner part of the Solar System would not be until 1992.

When the comet failed to turn up in the early 1980s, Marsden (a theorist who had no direct access to major telescopes) tried to encourage his professional observer colleagues to carry out a search for it in 1992. But he failed to arouse their interest, even though the Perseids in 1991 provided an unusually spectacular display visible from Japan – a hint, in Marsden's view, that the comet itself was approaching. It was left to amateur astronomers to keep a look out for the parent of the Perseids, and one of them did indeed find it, just as Marsden had anticipated.

Once the comet was found, of course, the professionals took notice and turned their telescopes on to it. One reason for their interest was the realization that on its next pass through the inner Solar System the comet would indeed come close to the Earth. But before they could calculate an improved orbit from their new observations, Marsden and his colleagues had done the trick by searching back through the old records in the light of the confirmation that Swift–Tuttle and the comet of 1737 were indeed one and the same. The earliest recorded observation of what is now known to be Swift–Tuttle was in 68 BC, and with data going back more than 2,000 years the astronomers were able to calculate that the comet is locked in a gravitational embrace with Jupiter. Jupiter itself orbits the Sun exactly 11 times for each single orbit of the comet, keeping it in a very stable and predictable orbit (it is still a mystery why the 1862 observations gave the wrong orbit, but that mystery is never likely to be solved; probably, somebody simply made a mistake).

Swift–Tuttle definitely will not hit the Earth in 2126, but because of this locking of its own orbit to Jupiter it will continue to make close approaches to the Earth from time to time for at least the next 10,000 years. 'The dates of those approaches cannot be known,' Marsden told us, but 'Swift–Tuttle is the largest threat known to mankind that will continue to be a threat for the longest time in the future'.

If you are inclined to worry about such things, there is a startling array of possible suggestions as to what to do about them. As well as the obvious value of keeping an eye out for objects in space that are in orbits that may intersect with the Earth, extremist suggestions as to what to do have included the idea of building laser guns on the Moon to blast incoming debris, and establishing an orbiting fleet of unmanned nuclear missiles, ready to blow dangerous asteroids to pieces. This over-enthusiasm, in some quarters in the early 1990s, for the idea of a space defence was a direct result of the end of the cold war – the military industrial complex, particularly in the United States, saw the prospects of funding for 'Star Wars' type defence programmes disappearing, and latched on to 'the threat from space' as a means of persuading government to part with tax dollars to keep the weapons manufacturers in business. But behind the hype there was – and is – genuine cause for concern, even if there is no need to react on such a dramatic scale.

The Spacewatch telescope survey itself, for example, has found that the number of small objects (with diameters less than about 50 metres) that cross the Earth's orbit may be as much as a hundred times greater than had been thought prior to the 1990s. When these pieces of cosmic debris strike the Earth's atmosphere, most of them burn up long before they can reach the ground, exploding at high altitude with an energy of up to about 10 megatonnes of TNT. But, according to Christopher Chyba, of the NASA Goddard Space Flight Center in Greenbelt, Maryland, the statistics suggest that six per cent or so of the objects found by Spacewatch ought to be rich in iron, and could therefore penetrate to the ground, producing, on average, one 20 kilotonne explosion every five years and one 700 kilotonne explosion every 350 years. The evidence suggests that many of these small objects are associated with the Earth's own mini-asteroid belt; objects which may have been perturbed into orbits similar to that of the Earth by the gravitational influence of Jupiter, or which may possibly represent debris blasted out of the Moon by major impacts long ago, which has stayed in much the same orbit around the Sun as the Earth–Moon system itself.

Whatever their origin, there are at least 200 known so-called near-Earth objects (NEOs), many of which more or less share the Earth's orbit around the Sun, and which range in size from 10 to 50 metres across. With one or two new NEOs being discovered each month, by the time you read these words the number known may well be above 500. It is estimated that 2,000 asteroids, each more than a kilometre across, are in Earth-crossing orbits, accompanied by a million objects between 0.1 km and 1 km across and at least 500 million smaller objects (down to 10 metres across).

So the threat is certainly real. But how do you measure such threats? How do you decide what response is cost effective? The best analysis, so far, of the risks involved has been carried out by Clark Chapman, of the Planetary Science Institute in Tucson, Arizona, and David Morrison, of the NASA Ames Research Center at Moffett Field, in California.[3] It has since provided the basis for all attempts at deciding how best to combat the threat from space.

Chapman and Morrison drew on several studies carried out during the 1980s and early 1990s, most of which had been inspired, directly or indirectly, by the realization that the Earth had suffered a major cosmic impact at the end of the Cretaceous. The bottom line of their calculation is that there is a 1 in 10,000 chance that an object bigger than 2 km across will strike the Earth in the next century, causing major disruption of the world's ecosystem and killing a large proportion of the human population of the planet. The risk is very small but, using the actuarial approach of life insurance companies, the number of people potentially affected is very large, and in statistical terms that puts the risk to any individual human being surprisingly high up the scale of things to worry about.

For any present-day citizen of the USA, the chance of being killed by the after effects of a cosmic impact are about 1 in 20,000. This is bigger than the risk of being killed in any other kind of natural disaster, and is exactly equal to the risk of being killed in an airplane accident on a commercial passenger flight, something that very many people take seriously as a risk to worry about. It compares favourably with the chance of being killed in a motor vehicle accident of 1 in 100, and with the danger of accidental electrocution of 1 in 5,000, but unfavourably with the chance of being killed by a tornado (1 in 60,000), and with the risk of death by food poisoning (just one in three million). The risk of death from cosmic impact is greater than the *combined* risk of death from tornadoes, hurricanes, earthquakes, forest fires and volcanic eruptions.

The point is, as Chapman and Morrison put it, that 'impacts are an extreme case of a low-probability/high-consequence hazard'. In his book *Exploring Planetary Worlds*, Morrison spells out how the odds are calculated. If an impact big enough to cause a global crop failure happens once every 375,000 years, and you have a 1 in 4 chance of dying if such an impact occurs in your lifetime, then the annual risk of death from this cause is 1 in (3 × 375,000), which is 1 in 1.5 million. If the average human lifetime is 75 years, the overall chance of being killed in this way is 75 times greater – 75 in 1.5 million, or 1 in 20,000.

The chance of anyone at all being affected in this way is indeed small, but if one person is affected by a major cosmic impact then literally billions of other people will suffer as well. This changes the way you should think about the risk. For example, many cautious people buy cheap 'one-off' life insurance policies every time they fly, so that if that 1 in 20,000 risk comes up, their dependents will, at least, have financial security. But there is no point in taking out comparable life insurance against the risk of death in a major cosmic impact, because if that event happens you probably won't leave any surviving dependents, and in any case there will be no insurance company left to pay up. The only kind of 'life insurance' against a major cosmic impact is to take global precautions, and the cost of those precautions should be seen in exactly those terms – as life insurance for billions of people. Divide the cost of the precautions by the number of people in the world (or even in the USA and Europe) and you have a sensible figure for the cost of the insurance in personal terms. This makes current proposals, to keep a proper watch out in our neighbourhood of space, put forward by astronomers look very cost effective – certainly better value for money than those one-off airline flight insurance premiums.

The chance of any one person being killed by a major comet impact in any one year is only about one in two million. But this still means that, on average, 2,700 people are killed by large comets each year, 390 of them in the developed world alone. Putting this in a slightly different perspective, the cost of road safety measures in Britain in the mid-1990s works out at just over £800,000 for each life that is saved. If we cared as much about the comet and asteroid hazard as we did about road safety, that would imply that the developed world ought to be spending nearly $500 million a year to prevent cosmic impacts. The fact that nobody is killed by such an impact for many years, and then many people are killed at once, does not alter the argument. There are long stretches of road where nobody is killed for many years, and no big accidents occur – but that is not a good argument for saving costs by not bothering to maintain safety barriers and decent lighting on those stretches of road on a year-by-year basis.

One of the points which Chapman and Morrison, among others, stress is that it is indeed comets that we should worry about most. Because comets come in to our part of the Solar System from much farther out, falling in and accelerating in the Sun's gravitational grip as they do so, by the time they cross the orbit of the Earth they are travelling much faster than asteroids, and so carry more kinetic energy. Size for size, an incoming comet will do more damage to the Earth than an asteroid. At the same time, whereas asteroids are usually found in well-established orbits around the Sun, and can be monitored for many orbits while the potential for future risk is assessed, 'new' comets (as many as ten a year) arrive unpredictably from the outer part of the Solar System. This, combined with their high speed, makes the time available to identify the risk and react to it much smaller than for a typical large asteroid.

It is now clear that a fireball with the energy equivalent of the Hiroshima bomb is produced high in the atmosphere about once a year by minor cosmic impacts. Although a sobering indication of the amount of cosmic debris around, these small impacts have no direct effect on the surface of the Earth. Even the kind of events we have just mentioned, producing local explosions on a Tunguska scale every few hundred years, are not the kind of thing that the spacewatchers worry about. If we are worrying about the risk of truly global disaster, everything up to the scale of explosions releasing energy equivalent to 1,000 megatonnes of TNT, even the kind of impact we discussed in the previous chapter, can be regarded as 'local'. Such events can be expected about once every 10,000 years, so anyone who lives to be 100 years old has a one per cent chance of being alive when an object big enough to produce such an explosion strikes the Earth. Devastation might extend over an area of tens of thousands of square kilometres, millions of people might be killed, and the resulting relatively minor cosmic winter effect might disrupt civilization worldwide for decades. Remember that a 1 °C drop in the average temperature of the world would wipe out commercial production of wheat in Canada, while a 3 °C drop in mean temperatures would shift the northern limit of corn production in the US to southern Iowa. But this would still be a merely local problem.

The threshold for a 'global impact catastrophe' is arbitrarily defined by Chapman and Morrison as one which leads to the death of a quarter of the human population of our planet. To do this requires an impact big enough to cause a cosmic winter so severe that temperatures worldwide fall by 10 °C for months or years, bringing killing frosts to mid-latitudes even in summer and destroying any prospect of growing food crops. The disaster would also, of course, be accompanied by all the usual associated effects that we discussed in connection with the death of the dinosaurs – acid rain, destruction of the ozone layer, and so on. To cross this threshold for global disaster would only require the impact of a stony meteorite 1.5 km across, producing an explosion equivalent to 200,000 megatonnes of TNT. With an average world population density of ten people per square kilometre, Chapman and Morrison calculate that 'only' some three million people would be killed by the direct effects of an impact just on the threshold for global catastrophe; while the number of indirect casualties, 1.5 billion people, would be 500 times greater. This is because only 0.1 per cent of the Earth's surface is directly affected by the blast in such an event, but the entire surface of the planet is affected by its consequences.

As we have mentioned, there is a 1 in 10,000 chance that such a disaster will occur in the next 100 years. There is also a 1 in 100 chance that an impact equivalent to 100 Tunguska Events taking place together will occur in the next 100 years, releasing the energy equivalent of 1,000 megatonnes of TNT.

The sensible response to this threat – ignoring the wild-eyed enthusiasm of a few people for putting huge nuclear arsenals into space 'just in case' – has come from a NASA workshop which in 1992 proposed the so-called Spaceguard Survey, a scheme to monitor near-Earth space for hazards using a purpose-built network of telescopes.[4] The Spaceguard plan would require building at least six (and possibly eight) new ground-based telescopes, each with an aperture of 2.5 metres. Although by no means the largest telescopes on Earth, this is hardly small beer by astronomical standards – until quite recently, the largest telescope operating on Earth was the famous 200-inch Hale telescope on Mount Palomar, in California, which has a main mirror with a diameter of 5 metres, just twice as big as the proposed Spaceguard telescopes.

With a start-up cost of some $50 million to build the necessary telescopes, and running costs of just $10 million per year, Spaceguard could identify all the potentially dangerous asteroids in the inner Solar System over a period of just 20 or 30 years, and could also give a few months' warning of the arrival of a threatening comet. Any warning is better than none, of course, but this short timescale for comet warnings is definitely the Achilles heel in the basic Spaceguard proposal. Even so, the cost represents much less than a dollar per person for the population of the United States alone, which compares very favourably with the cost of most kinds of life insurance. Rather surprisingly, though, NASA has in recent years allocated no more than $500,000 to modest spacewatching efforts in any one year, and nobody else is devoting any serious money to the problem at all (although there have been suggestions that the US Air Force might contribute one or more of its existing telescopes, at present used to track satellites in space, to get Spaceguard up and running).

Morrison has pointed out that at the time of writing (May 1995) there are less than a dozen people in the whole world who are searching for NEOs – 'fewer people', as he puts it, 'than it takes to run a single McDonald's'. The longest-running of these surveys was set up by Eugene Shoemaker and Eleanor Helin, of CalTech, in 1973. For four or five nights each month, they photograph large areas of the sky using long exposures with a telescope with a wide field of view, known as a Schmidt camera, at the Palomar Observatory. Then they scan the photographs for the distinctive trail left by asteroids moving across the field of view. Charles Kowal, who found Chiron, is another regular asteroid watcher, and we have already mentioned the Spacewatch program. On a particular night, this obtains repeated images of the same part of the sky using an electronic system called a charge coupled device (CCD) attached to a conventional telescope. A computer then electronically 'subtracts' one image from another, so that any objects which stay in the same place (that is, stars) disappear, while anything that has moved between the two observations (such as an asteroid) shows up as a double image in the subtracted picture. The technology is neat and effective. But overall this still seems a ridiculously modest investment in early warning, in view of the size of the potential damage that an impact could cause.

At least, that is our conclusion. There are people who disagree. Carl Sagan, of Cornell University, and Steven Ostro, from the NASA Jet Propulsion Laboratory, pointed out in 1994 that, averaging the risk out on an annual basis, the number of deaths per year from cosmic impact is about 3,000, which compares with, they point out, *three million* deaths per year resulting from smoking tobacco. If people keep smoking tobacco over the next thousand years, twice as many people will die as a result than would die in a threshold global disaster impact. But the crucial difference, we suggest, is that in the impact disaster they all die at once, and civilization collapses. Indeed, we could turn Sagan and Ostro's argument on its head, and use it to indicate the seriousness of a threshold impact event by pointing out that it would kill as many people in one go as are likely to die from smoking tobacco (at present rates) in five hundred years!

It is even possible that the original cost estimates for the proposed Spaceguard programme are too high. The relevant technology has been advancing rapidly in the 1990s and, as we have mentioned, astronomers now routinely use CCDs to scan the heavens on a large scale. These techniques have developed in leaps and bounds over the past few years, partly because astronomers have been carrying out enormous surveys involving imaging millions of stars each night. They are searching for 'flickering' in a very few of those stars caused by the gravitational influence of dark objects (which are thought to make up most of the mass of a galaxy like our Milky Way) passing in front of them. For many purposes, CCDs replace photographic surveys, and have the advantage that the images they produce can be monitored automatically, almost in 'real time', by computers. With the aid of CCDs and modern computers, the Spaceguard survey might now be feasible (according to Ted Bowell, an astronomer at the Lowell Observatory) for less money in less time; perhaps taking only ten years, and certainly costing no more than a single space mission.

Even if Spaceguard, or something similar, were set up, what would we do if and when we found a large object on a collision course with Earth? A summary of discussions of this problem among experts at several recent workshops was reported to the 1994 meeting of the American Association for the Advancement of Science by Gregory Canavan, of the Los Alamos National Laboratory in New Mexico. Spaceguard would identify about a thousand NEOs a month, most of which would quickly be identified as having non-threatening orbits. But many would have to be studied in more detail to check just how close they were likely to come to the Earth. In order to pin down the orbit of such an object precisely, several large radar systems would be needed (in both the northern and southern hemispheres). These could confirm which of the objects really did pose a threat, and provide invaluable help in tracking any probe sent up to the object to the point of rendezvous.

Most of the studies carried out so far assume, as we have suggested, that the threatening object would be an asteroid, and that there would be time to study it carefully over several orbits before any action need be taken. A lightweight spaceprobe could be sent on such a mission, taking a few months to reach the object of interest, using standard launch vehicles. Such a probe could measure the exact size of the threatening object, and determine its composition, so that it would be possible to calculate how easy it would be to deflect or disrupt the object.

Blowing the object apart has a certain appeal in some quarters, but might not be the best thing to do. If an asteroid on an intercept orbit with the Earth is blown into a hundred pieces, the hundred pieces will each still be on an intercept orbit with the Earth. Although the resulting shotgun effect of their multiple impacts might not be quite so bad as a single huge impact, it would still hardly be desirable. If the object were small enough and fragile enough, it might be deemed worthwhile to break it into pieces small enough to burn up in the Earth's atmosphere.

Although the usual image from science fiction movies is of a fleet of nuclear-tipped missiles being sent out to do such a job, that would hardly be necessary. The best approach might be simply to use the kinetic energy of the object to bring about its own downfall, by manouvering a large interceptor rocket into its path. If the impact velocity was only 30 km per second, the energy released in the collision would be a hundred times as much as would be released by an amount of high explosive equal to the mass of the rocket. Such an impact could also be used to deflect the incoming object onto a new trajectory, and deflection is by far the favoured option in most scenarios of this kind. It only takes a modest sideways nudge to push an object into an orbit which takes it clear of the Earth, and the earlier the nudge can be made the less effort is needed to do the job.

Collisions, or even nuclear explosions, could do the job, or rocket thrusters could be installed on the object itself. This is a particularly neat idea in the case of comets, because the material from the comet itself (the ice from its nucleus) could be used as the reaction mass for the rockets, steadily pushing it into a new orbit.[5] There is even a way to do the same trick for a rocky or iron asteroid, installing a device called a 'mass driver' which would work by throwing lumps of rock (or iron) away into space. Obeying Newton's law (action and reaction are equal and opposite), every lump thrown off the asteroid would give it a nudge in the opposite direction – just as an ice skater who tries to throw a heavy object away will find that he or she slides off in the opposite direction. But as yet these ideas are still in the realm of science fiction.

Canavan summed up the possibilities for existing technology, using explosives or impacts to do the deflection and assuming a maximum deliverable payload of five tonnes to deep space, in his AAAS talk. The detonation of a single one kilotonne warhead in the right place (buried beneath the surface of an asteroid) would be enough to nudge an asteroid 100 metres or so in diameter sideways at a rate of one centimetre per second – which, given enough seconds, could take it as far from the Earth as you like. Much bigger warheads are, of course, already available, and a single one megatonne blast would shift a kilometre-sized asteroid by the same amount. With a hundred years to react, a suitably large subsurface nuclear explosion (or a series of smaller explosions) would be sufficient to deflect even an NEO 100 km in diameter from an intercept trajectory with the Earth. With ten years warning a similar explosion could deflect an object 30 km across. Thomas Ahrens, of CalTech, has coined a neat term for the use of nuclear explosives in this way – 'weapons of mass protection'. Kinetic energy – simply dumping the five tonnes in the right place in front of the NEO – would be enough to deflect a two kilometre diameter object given 10 years' warning.

Whether it is worth making the effort depends on how much damage the incoming object is likely to do. It would be nice to think that it is always worth saving lives, but at some point the cost of insurance becomes prohibitive. Even in everyday terms, there are some risks that we do not insure against, because we think we have better uses for the money. So we need to know how much the impacts would cost us, before we decide how much we should spend to prevent them. The figures reported by Canavan suggest that the average cost of impacts by small metallic NEOs is about $10 million per year, while impacts from objects up to about 2 km across cost $100 million a year. This doesn't mean, of course, that this is the actual cost of impacts year by year, but rather that an impact causing damage worth $10,000 million occurs once every hundred years. For impacts above the threshold for global catastrophe, the guesstimate is that the cost is equivalent to a loss of $500 million per year in economic terms. On that basis, he concludes, it would be worth investing up to $100 million per year from the global economy in order to provide an effective defence against incoming cosmic objects with diameters up to about 8 km; and 'defenses that detect NEOs many orbits prior to impact are cost effective for large NEOs'.

There is one other great advantage of this kind of global life insurance – if we are only concerned about asteroids in predictable, long-term orbits, the defence system itself (the rockets and other hardware) need not be built until the threat has been positively identified. Appropriate systems can be designed, and prototypes tested, but the mass production of the weapons of mass protection need only take place when it is known that a threatening cosmic object is on the way.

The first prototypes of some of the equipment are already being tested in space – a tiny American spaceprobe called Clementine, launched in 1994, made a rendezvous in 1995 with an asteroid known as Geographos, which is 20 km across and was at the time just twice as far away from us as the Moon. There are also plans for a NASA probe to be launched in 1996 on a trajectory which will eventually (late in 1998) place it in orbit around Eros. Clementine, in particular, was intended as a test of 'Star Wars' technology, including advanced sensors, computer systems and image processors. It has been paid for out of the budget of the Ballistic Missile Defense Organization (BMDO, the latest official name for the US Star Wars programme). But that is exactly, as we have seen, the technology required to deal with incoming asteroids.

Spaceguard would require at least three telescopes in each hemisphere to look right around the sky every 24 hours as the Earth rotates. This is necessary because objects that are relatively close to the Earth move rapidly across the sky, and it is no good looking in one part of the sky one night and another patch of sky the next night, because the objects of interest may have moved right across the field of view of the instrument and be somewhere else by the time you look at a particular patch of sky. Unfortunately, there is a trade-off between being able to look at a large patch of the sky (using a wide angle telescope) and being able to identify very faint (and therefore distant) objects, using a narrow-angle telescope. With the size of instruments envisaged in the original proposal, long-period comets would only be identified as they came well within the orbit of Jupiter: probably only a few months – a year at most – before any potential impact with the Earth. The kind of response outlined above would require a much earlier early warning than this.

Tentative plans for a 'Spaceguard 2' using larger telescopes (with, perhaps, twice the aperture of the instruments used in 'Spaceguard 1', but a narrower field of view) show how this problem might begin to be tackled. Because these larger telescopes would be specifically dedicated to looking for faint comets beyond Jupiter, in the giant planet region where they spend many years, only one instrument would be needed in each hemisphere to scan the skies on a more leisurely basis. At such distances, comets are scarcely moving across the line of sight, but are falling in towards the Sun, which means more or less towards the Earth. So they do not shift much across the sky from night to night, or even from week to week – and one telescope can take its time to look around the sky in order to locate them.

The realisation of any such plans lies far into the future, and even Spaceguard 1 has not yet, as we have seen, got the go ahead. As of early 1995, the best that seemed likely to happen was an even cheaper, scaled down version of the

Spaceguard Survey, which would do nothing at all to address the glaring weak point in what was otherwise a well conceived idea, the problem of long-period comets.

This is the major snag about Spaceguard 1, even if it does get the go-ahead in any form. Because almost all asteroids orbit in the same plane around the Sun as the orbits of the planets, the standard 'search pattern' for NEOs usually envisaged for Spaceguard concentrates on this region of the Solar System. But long-period comets (LPCs) can come in from a much wider range of angles, and this means that the Spaceguard telescopes might simply be looking in the wrong direction to spot them, even after they pass within the orbit of Jupiter. The obvious immediate resolution of this difficulty is to extend the search to higher astronomical latitudes – but that would involve building more telescopes, at greater expense. To do the job properly would require 18 telescopes instead of six, and it seems highly unlikely that this will be done in the foreseeable future.

In their contribution to the volume *Hazards Due to Comets and Asteroids* (edited by Tom Gehrels), Brian Marsden and Duncan Steel point out that, if surveys for objects in orbits that intersect with the Earth are restricted to the search pattern envisaged in the Spaceguard proposal, as many as a quarter of potentially impacting long-period comets would never be observed (at least, not unless and until some lucky amateur spotted them). They suggest that 'discovery 100 days before a prospective impact is the minimum acceptable leadtime' for action to be taken. This seems wildly optimistic in our view, and would require that some sort of immediate response capability, presumably in the form of nuclear-tipped spaceprobes, was kept on permanent standby for such an eventuality. But even taking this at face value, as Marsden and Steel go on to say, 'it is sobering to realize that only 80 comets, or less than 12 per cent of the total set of known LPCs, have been discovered at a time interval 100 days before perihelion' (which is equivalent to finding an Earth-impactor at least 100 days before it reaches us).

If comet impact is the way the world as we know it will end, as things stand today we will probably have less than three months' warning of our doom. Perhaps it really is time to take the threat seriously, and apply the technology which has been shown to be effective by the Spacewatch project on a much larger scale. The alternative is to bury our heads in the sand, hope that it doesn't happen in our own lifetimes, and save every person on the planet the princely sum of one fifth of one US cent each year. If you think it would be worth paying one cent every five years to insure yourself against the *combined* risk of tornadoes, hurricanes, earthquakes, forest fires and volcanic eruptions, then maybe you will agree with us that it is worth spending a similar sum to provide protection against cosmic impacts.

Notes

1 The telescope dates back to the early 1920s, but was adapted to carry out an automatic scan of the sky for passing asteroids at the beginning of the 1990s; it is run by a team at the University of Arizona, headed by Tom Gehrels.

2 Volume 78, p. 654.

3 *Nature*, 6 January, 1994 (volume 367, p. 33).

4 The name 'Spaceguard' was deliberately lifted from the name of a fictional spacewatching service mentioned in Arthur C. Clarke's novel *Rendezvous with Rama* (Gollancz, London, 1973). Clarke later

repaid the compliment by writing a novel about cosmic impacts, *The Hammer of God* (Bantam, New York, 1993).
5 It is such a neat idea, indeed, that one of us (JG) used it in an SF novel, *Double Planet*.

FURTHER READING

Slightly more technical books are marked with an asterisk; the rest are accessible at about the level of this book.

John Brandt & Robert Chapman, *Rendezvous in Space*, W. H. Freeman, New York (1992).

Nigel Calder, *The Comet is Coming!*, BBC Publications, London (1980).

*S. V. M. Clube (ed.), *Catastrophes and Evolution*, Cambridge University Press, Cambridge (1989).

Victor Clube & Bill Napier, *The Cosmic Serpent*, Faber & Faber, London (1982).

Victor Clube & Bill Napier, *The Cosmic Winter*, Blackwell, Oxford (1990).

John Davies, *Cosmic Impact*, Fourth Estate, London (1986).

Stephen Edberg & David Levy, *Observing comets, asteroids, meteors, and the zodical light*, Cambridge University Press, Cambridge (1994).

Peter Francis, *The Planets*, Pelican, London (1981).

*Tom Gehrels (ed.), *Hazards Due to Comets and Asteroids*, University of Arizona Press, Tucson (1995).

William Glen (ed.), *The Mass-Extinction Debates*, Stanford University Press, Stanford (1994).

John Gribbin, *In the Beginning*, Penguin, London (1994).

John Gribbin, *Companion to the Cosmos*, Weidenfeld & Nicolson, London (1996).

John Gribbin & Marcus Chown, *Double Planet*, Gollancz, London (1988).

Mary Gribbin & John Gribbin, *Being Human*, J. M. Dent, London (1993).

*Jean Grove, *The Little Ice Age*, Methuen, London (1988).

Nigel Henbest, *The Planets*, Viking, London (1992).

*Fred Hoyle, *The Cosmogony of the Solar System*, University College Cardiff Press, Cardiff (1978).

Fred Hoyle, *Ice*, Hutchinson, London (1981).

John Imbrie & Katherine Palmer Imbrie, *Ice Ages*, Enslow, New York (1978).

William J. Kaufmann, *Exploration of the Solar System*, Macmillan, New York (1978).

*E. L. Krinov, *Giant Meteorites*, Pergamon, London (1966).

Hubert Lamb, *Climate, History and the Modern World*, Methuen, London (1982).

Peter Lancaster-Brown, *Halley & His Comet*, Blandford, Poole (1985).

*Kenneth Lang & Charles Whitney, *Wanderers in Space*, Cambridge University Press, Cambridge (1991).

G. J. H. McCall, *Meteorites and their Origins*, David & Charles, Newton Abbot (1973).

*G. J. H. McCall, *Meteorite Craters*, Dowden, Hutchinson & Ross, Stroudsburg, Pennsylvania (1977).

*David Morrison (ed.), *The Spaceguard Survey*, Jet Propulsion Laboratory/California Institute of Technology, Pasadena (1992).

David Morrison, *Exploring Planetary Worlds*, Scientific American/W. H. Freeman, New York (1993).

Richard Muller, *Nemesis: The Death Star*, Heinemann, London (1988).

Richard Norton, *Rocks from Space*, Mountain Press Publishing, Missoula (1994).

Plato, *Timaeus and Critias* (trans. H. D. F. Lee), Penguin, London (1971).

David M. Raup, *The Nemesis Affair*, Norton, New York (1986).

David M. Raup, *Extinction*, Norton, New York (1991).

Carl Sagan & Ann Druyan, *Comet*, Guild Publishing, London (1985).

*Roman Smoluchowski, John Bahcall & Mildred Matthews (eds.), *The Galaxy and the Solar System*, University of Arizona Press, Tucson (1986).

Steven M. Stanley, *Extinction*, Scientific American/W. H. Freeman, New York (1987).

Duncan Steel, *Rogue Asteroids and Doomsday Comets*, Wiley, Chichester (1995).

John Stirling (ed.), *The Bible*, British & Foreign Bible Society/Oxford University Press, Oxford (1954).

Fred Whipple, *The Mystery of Comets*, Smithsonian Institute Press, Washington DC (1985).

Michael White & John Gribbin, *Darwin: A life in science*, Simon & Schuster, London (1995).

INDEX

Deccan Traps, 21
deforestation, 51, 56, 211
Delporte, Eugene, 97
Devonian period,
    extinctions, 9,10
diamond dust, 151-4, 158
diamonds, K-T boundary,
    19-20
dinosaurs:
    extinction, 1-23, 38-9,
        161
    fill available ecological
        niches, 12
    period of, 13
    survivors of the K-T
        catastrophe, 13
doom, comets as
    harbingers of, 205
dust, 47-8
    annual rate of deposit,
        81
    comets, 110, 113
    diamond dust, 152-4,
        158
    ice cores, 212, 218
    little ice ages, 215-17,
        218-19
    plant formation, 66
    zodiacal, 156-9, 160,
        161, 163
Egypt, ancient, 200, 210
Encke, Johann Franz, 101,
    134

Encke's Comet:
    debris, 136-40, 203,
        213, 219
    and devastation in
        Britain, 213
    discovery, 134-5
    fire festivals and, 211
    nearness to Earth, 136
    orbit, 101-2, 136, 157
    and the Taurid stream,
        136-40, 203, 211, 219
energy, impact releases, 3
Enever, Joe, 3-7
English Civil War, 218
Epimethus, 86
Eros, 97, 240
ES1, 223
Europa, 84-5
evaporites, 34
Evenki, 52
evolution, 39-40
explosions:
    cosmic debris, 227
    fireballs, 224
    scale of, 43-5
    size of, 227
explosives, deflecting
    asteroids, 237-9
extinctions, 7-10
    Big Five, 9
    comet impacts and, 134
    continental drift, 8
    cosmic impact and, 1-22

254